Lecture Notes in Statistics 113

Edited by P. Bickel, P. Diggle, S. Fienberg, K. Krickeberg, I. Olkin, N. Wermuth, S. Zeger

Springer
New York
Berlin
Heidelberg
Barcelona
Budapest
Hong Kong
London
Milan
Paris
Santa Clara
Singapore
Tokyo

Rainer Schwabe

Optimum Designs for Multi-Factor Models

Springer

Rainer Schwabe
Fachbereich Mathematik und Informatik
Freie Universitat Berlin
Arnimallee 2-6
14195 Berlin
Germany

CIP data available.
Printed on acid-free paper.

Camera ready copy provided by the author.
Printed and bound by Braun-Brumfield, Ann Arbor, MI.
Printed in the United States of America.

9 8 7 6 5 4 3 2 1

ISBN 0-387-94745-0 Springer-Verlag New York Berlin Heidelberg SPIN 10490817

Preface

The design of experiments in large systems has attracted a growing interest during the last years due to the needs in technical and industrial applications, mainly connected with the philosophy of quality design. An appropriate desription of such a systems can be given by a multi-factor model in which the outcome of an experiment is influenced by a number of factors which may interact in a suitable way. The intention of the present monograph is the presentation of the theory and the methodology for the generation of optimum designs in important, large classes of such multi-factor models which are specified by their interactions structures.

The present notes emerged from a larger research project on efficient experiments in industrial applications which ran for several years at the Free University of Berlin, and the material covered has been successfully used for a series of lectures held at Uppsala University in 1994. My hope is that the results presented here are stimulating for further research in this area.

I am indebted to all my friends and colleagues who encouraged me to finish this work. My particular thanks are to V. Kurotschka for our continuing discussions on the philosophy of the design of experiments and to W. Wierich for introducing me to the secrets of the theory at an early stage. Th. Müller-Gronbach did an efficient job in proof-reading on a preliminary version of these notes, and some remaining misprints were detected by H. Großmann and A. Gut. I very much appreciated the help of A. Behm in type-setting major parts of the manuscript and the patience of the corresponding editors M. Gilchrist and J. Kimmel of the *Lecture Notes in Statistics* series while several deadlines passed by. Finally, this research was generously supported by the *Deutsche Forschungsgemeinschaft.*

Rainer Schwabe, *Berlin, January 1996*

Preface

Table of Contents

Part I

General Concepts

The design of an experiments aims at an efficient choice of the experimental conditions in order to optimize the performance of the statistical inference which will be based on the observations to be obtained. This search for optimum experimental settings has give rise to many theoretical investigations. The first paper which was explicitly devoted to an efficent experimental design was written by SMITH (1918) at a time when no general concepts had been developed, yet. It was, then, FISHER (1935) who spread general ideas of experimental design. These ideas were strongly connected with the concept of linear models which is sketched in Subsection 1.1. As in many other fields of statistics WALD (1943) gave a first concise formulation of the optimization problem. In the fifties a broad investigation of experimental design problems started (see e.g. ELFVING (1952), EHRENFELD (1955), GUEST (1958) and HOEL (1958), besides others). One of the outstanding personalities in the design of experiments became KIEFER who initiated a whole new theory with his introduction of generalized (or approximate) designs (cf KIEFER and WOLFOWITZ (1959)) which will be described in Subsection 1.2.

It was also KIEFER (1959) who first discussed a variety of different optimality criteria for experimental designs, and KIEFER and WOLFOWITZ (1960) established that the D-criterion of generalized variance and the G-criterion of minimaxity within the design region are optimized simultaneously by the same design. This result is, now, known as the KIEFER-WOLFOWITZ eqivalence theorem, and it was extended to the A-criterion of minimum EUCLIDEan norm by FEDOROV (1971) and to general equivalence theorems by WHITTLE (1973) and KIEFER (1974). These, and some other results are exhibited in Section 2. For the general theory and further developments in the field of optimum experimental designs we refer to a series of monographs by FEDOROV (1972), BANDEMER et al. (1977), KRAFFT (1978), BANDEMER and NÄTHER (1980), SILVEY (1980), ERMAKOV et al. (1983), PÁZMAN (1986), ERMAKOV and ZHIGLYAVSKY (1987), SHAH and SINHA (1989), ATKINSON and DONEV (1992), PUKELSHEIM (1993) and BANDEMER and BELLMANN (1994). We also want to recommend the collection of KIEFER's (1985) papers on experimental design and the review articles by ATKINSON (1982–1995).

To find an optimum design standard methods like invariance and orthogonalization can be used, which date back, at least, to KIEFER (1959) and KIEFER and WOLFOWITZ (1959). In many situations these methods which are presented in Section 3 sufficiently reduce the design problem and lead to explicit construction of an optimum design.

1 Foundations

In every experimental situation the outcome of an experiment depends on a number of factors of influence lie temperature, pressure, different treatments or varieties. This dependence can be described by a functional relationship, the response function μ, which quantifies the effect of the particular experimental condition $t = (t_1, ..., t_K)$. The observation X of an experiment is subject to a random error Z. Hence, an experimental situation will be formalized by the relationship $X(t) = \mu(t) + Z(t)$. As the response function μ describes the mean outcome of the experiment the observation has to be centered and it is natural to require that $E(X(t)) = \mu(t)$ or, equivalently, that the error Z has zero expectation, $E(Z(t)) = 0$. The distribution of the random error may depend on the experimental conditions. In case of complete ignorance on the structure of the response function it will be impossible to make any inference. Therefore, we will consider the rather general situation in which the response is described by a linear model in which the response function μ can be finitely parametrized in a linear way as introduced in Subsection 1.1. The performance of the statistical inference depends on the experimental conditions for the different observations, and it is one of the challenging tasks to design an experiment in such a way that the outcome is most reliable. This concept will be explained in Subsection 1.2.

1.1 The Linear Model

In an underlying experimental situation $X(t) = \mu(t) + Z(t)$ we will say that the response function μ can be parametrized in a linear way, if μ is known to be a linear combination of some known regression functions $a_1, ..., a_p$ on the set T of possible settings for the experimental condition t, $\mu(t) = \sum_{i=1}^{p} a_i(t)\beta_i$. Hence, the structure of the response function μ is known up to a finite number of parameters $\beta_1, ..., \beta_p$. For notational ease we write the parameters and regression functions as vectors $\beta = (\beta_i)_{i=1,...,p} \in \mathbb{R}^p$ and $a = (a_i)_{i=1,...,p} : T \rightarrow \mathbb{R}^p$, respectively. Hence, an experimental situation $X(t) = \mu(t) + Z(t)$ is described by a *linear model* if $\mu(t) = a(t)'\beta, t \in T$, (or equivalently $E(X(t)) = a(t)'\beta$) for some known regression function a.

The representation $\mu = a'\beta$ of the response function is not unique because any linear one-to-one transformation given by $\widetilde{a} = La$, for $L \in \mathbb{R}^{p\times p}$ non-singular, causes a reparametrization $\mu = \widetilde{a}'\widetilde{\beta}$ of the response function and the corresponding parameters β and $\widetilde{\beta}$ are linearly connected by $\widetilde{\beta} = L'^{-1}\beta$.

In most situations (if not explicitly stated the opposite) we assume a *minimum* parametrization, i. e. the components $a_1, ..., a_p$ of the regression function a are linearly independent on the design region T of possible experimental conditions. Then any two minimum linear parametrizations $\mu = a'\beta = \widetilde{a}'\widetilde{\beta}$ are connected by a unique non-singular matrix $L \in \mathbb{R}^{p\times p}$ as indicated above.

We want to emphasize that a linear model is completely specified by the linear parametrization of its response function μ. For illustrative purposes we start with two simple examples in which only one factor of influence is involved. In the first example the factor is qualitative, i. e. it can be adjusted to a finite number of levels which corresponds to a qualitative property like different treatments.

Example 1.1 *One-way layout:*
The qualitative factor can be adjusted to I different levels and the parameters are the effects of each single level i, which will be denoted by α_i,

$$\mu(i) = \alpha_i \ ,$$

$i = 1, ..., I$. The dimension of the unknown parameter $\beta = (\alpha_i)_{i=1,...,I}$ equals $p = I$ and the regression functions are given by indicators, $a_j(i) = 1$ if $i = j$ resp. $a_j(i) = 0$ if $i \neq j$. Thus $a(i) = (a_j(i))_{j=1,...,I} = \mathbf{e}_{I,i}$ is a vector of length I with its ith entry equal to one and all other entries equal to zero. □

In the next model the factor of influence is quantitative like temperature, pressure, processing time etc. and ranges over a continuum of levels.

Example 1.2 *Polynomial regression:*
The response function of this linear model is a polynomial in t of given degree $p-1$,

$$\mu(t) = \sum\nolimits_{i=1}^{p} t^{i-1}\beta_i \ ,$$

$t \in T \subseteq \mathbb{R}$. Then the regression function a is given by $a(t) = (1, t, t^2, ..., t^{p-1})'$. In particular, we will lay special interest in the models of linear regression ($p = 2$),

(i) $$\mu(t) = \beta_0 + \beta_1 t$$

and quadratic regression ($p = 3$),

(ii) $$\mu(t) = \beta_0 + \beta_1 t + \beta_2 t^2$$

(after a natural relabeling of the parameters). □

In most practical applications we are faced by more than one factor of influence and the complexity of the model is substantially higher. As the present work aims to reduce this complexity many categories of multi-factor examples will be considered later on. At this point only the seemingly simplest situation will be introduced,

Example 1.3 *Two-way layout:*
We consider two qualitative factors which may be adjusted to I resp. J different levels, hence $\mu : \{1, ..., I\} \times \{1, ..., J\} \to \mathbb{R}$. In this situation the response function is not self-evident as it is in Example 1.1 where only one qualitative factor is involved. Different interaction structures may occur,

(i) *Complete interactions:*

$$\mu(i, j) = \alpha_{ij} \ ,$$

$i = 1, ..., I$, $j = 1, ..., J$. In this case each level combination (i, j) results in a different effect (for a further treatment see Example 4.1).

(ii) *Without interactions:*

$$\mu(i, j) = \alpha_i^{(1)} + \alpha_j^{(2)} \ ,$$

$i = 1, ..., I$, $j = 1, ..., J$. In this situation the effects of the first and of the second factor are additive (see Example 5.1). Note that the present parametrization is not minimum (cf Example 3.7) and the model has to be treated more carefully. □

Usually there are several observations which follow the particular model $X_n(t^{(n)}) = \mu(t^{(n)}) + Z_n(t^{(n)}) = a(t^{(n)})'\beta + Z_n(t^{(n)})$, $n = 1, ..., N$, at different experimental conditions $t^{(n)}$ for the factors of influence. In matrix notation these relations can be written as $X = A\beta + Z$, where $X = (X_n(t^{(n)}))_{n=1,...,N}$ and $Z = (Z_n(t^{(n)}))_{n=1,...,N}$ are the vectors of observations and of random noise, respectively, and the matrix $A = (a(t^{(1)}), ..., a(t^{(N)}))' = (a_i(t^{(n)}))_{n=1,...,N}^{i=1,...,p} \in I\!R^{N\times p}$ describes the influence of the experimental conditions $t^{(1)}, ..., t^{(N)}$ in terms of a. Besides the natural centering property $E(Z) = 0$ we will make the common assumptions that the random errors are uncorrelated, $cov(Z_n(t^{(n)}), Z_m(t^{(m)})) = 0$ for $n \neq m$, and that they are equally dispersed (*homoscedastic*), $Var(Z_n(t^{(n)})) = \sigma^2$. This reads in matrix notation as $Cov(Z) = \sigma^2\mathbf{E}_N$, where $\mathbf{E}_N$ is the $N \times N$ identity matrix.

Under these assumptions we obtain by the well-known GAUSS-MARKOV theorem that $\widehat{\beta} = (A'A)^{-1}A'X$ is the best linear unbiased estimator of β, as long as $A'A$ is regular, and in this case $Cov(\widehat{\beta}) = \sigma^2(A'A)^{-1}$. Moreover, $\widehat{\beta}$ is the best unbiased estimator in case of GAUSSian noise. In case that $A'A$ is singular there is no linear unbiased estimator of the unknown parameter β.

Frequently, our interest is confined to a linear function ψ of the parameter vector which is defined by $\psi(\beta) = L_\psi\beta$. For example, in the two-way layout without interactions (Example 1.3) a typical linear aspect ψ is the subset of parameters corresponding to the first factor and the effects described by the parameters of the second factor are considered as additional block effects.

An s-dimensional linear aspect *of β is a linear function $\psi : I\!R^p \to I\!R^s$ defined by $\psi(\beta) = L_\psi\beta$, and $L_\psi \in I\!R^{s\times p}$ is the associated selection matrix.*

For linear aspects ψ we may weaken the assumption that $A'A$ is regular and replace it by the requirement that ψ is (linearly) estimable, i.e. there exists a linear estimator $\widehat{\psi} = \widehat{\psi}(X)$ of ψ which is unbiased, $E(\widehat{\psi}) = \psi$. Often it is much easier to check the equivalent algebraic condition that ψ is (linearly) identifiable, i.e. $A\beta = A\widetilde{\beta}$ implies $\psi(\beta) = \psi(\widetilde{\beta})$. Hence ψ has to be a linear function of $A\beta$, or in terms of the corresponding matrix, $L_\psi = \widetilde{M}A$ for some $s \times N$ matrix $\widetilde{M}$.

By the GAUSS-MARKOV theorem the best linear unbiased estimator for an identifiable linear aspect ψ is given by $\widehat{\psi} = L_\psi(A'A)^-A'X$, which is independent of the particular choice of the generalized inverse $(A'A)^-$ of the matrix $A'A$. (For the general concept of generalized inverses we refer to the summaries in RAO (1973), pp 24, and CHRISTENSEN (1987), pp 336, and to the extensive monograph by RAO and MITRA (1971).) If ψ is identifiable, then $Cov(\widehat{\psi}) = \sigma^2L_\psi(A'A)^-L'_\psi$ which is, again, indpendent of the particular choice of the generalized inverse of $A'A$. Now, by the identity $A(A'A)^-A'A = A$ (cf Lemma A.1) we obtain an alternative formulation for the linear identifiability,

A linear aspect $\psi : I\!R^p \to I\!R^s$ defined by $\psi(\beta) = L_\psi\beta$ is identifiable if and only if there exists a matrix $M \in I\!R^{s\times p}$ such that $L_\psi = MA'A$.

A linear aspect ψ is *minimum* if its components are linearly independent, i.e. if the corresponding matrix L_ψ has full row rank, $\text{rank}(L_\psi) = s$. (For further readings on linear models we refer to SEARLE (1971), RAO (1973), and CHRISTENSEN (1987) besides others.)

1.2 Designed Experiments

The factors of influence can usually be controled by the experimenter. Hence, the realizations $t^{(1)}, ..., t^{(N)}$ of the levels at which observations are to be obtained can be chosen from the region T of possible experimental conditions to improve the performance of the resulting inference.

A design *d of size N is a vector $(t^{(1)}, ..., t^{(N)})$ of possible level combinations $t^{(n)} \in T$, $n = 1, ..., N$, for the experimental situation. The set T of possible combinations is called the* design region.

As we assume that the underlying experimental situation is a linear model with response function $\mu = a'\beta$, $a : T \to I\!R^p$, the designed experiment is specified by $X_d = A_d\beta + Z_d$ for every design d of size N, where $X_d = (X_n(t^{(n)}))_{n=1,...,N}$ and $Z_d = (Z_n(t^{(n)}))_{n=1,...,N}$ are the vectors of observations and random noise, respectively, and $A_d = (a_i(t^{(n)}))_{n=1,...,N}^{i=1,...,p} = (a(t^{(1)}), ..., a(t^{(N)}))'$ is the *design matrix* of the design d. The set of all possible designs of size N is given by T^N.

The quality of the best linear unbiased estimator $\widehat{\beta}_d = (A_d'A_d)^{-1}A_d'X_d$ for β resp. $\widehat{\psi}_d = L_\psi(A_d'A_d)^{-}A_d'X_d$ for an identifiable linear aspect ψ with $\psi(\beta) = L_\psi\beta$ is usually measured in terms of the corresponding covariance matrix and depends on the underlying design d. Because of the relations $Cov(\widehat{\beta}_d) = \sigma^2(A_d'A_d)^{-1}$ resp. $Cov(\widehat{\psi}_d) = \sigma^2 L_\psi(A_d'A_d)^{-}L_\psi'$ for the covariance matrices the $p \times p$ matrix $A_d'A_d$ plays an important role.

$\mathbf{I}^{(N)}(d) = \frac{1}{N}A_d'A_d$ *is the (normalized)* information matrix *of the design d of size N.*

Hence, $Cov(\widehat{\beta}_d) = \frac{1}{N}\sigma^2\mathbf{I}^{(N)}(d)^{-1}$ and $Cov(\widehat{\psi}_d) = \frac{1}{N}\sigma^2 L_\psi\mathbf{I}^{(N)}(d)^{-}L_\psi'$. In case of GAUSSian random errors $\mathbf{I}^{(N)}(d)$ is indeed FISHER's information for the location parameters up to multiplicative constants. Note also that $\mathbf{I}^{(N)}(d) = \frac{1}{N}\sum_{n=1}^{N} a(t^{(n)})a(t^{(n)})'$ is positive-semidefinite.

A linear aspect $\psi : I\!R^p \to I\!R^s$ defined by $\psi(\beta) = L_\psi\beta$ is identifiable *under d if and only if there exists a matrix $M_\psi \in I\!R^{s\times p}$ such that $L_\psi = M_\psi\mathbf{I}^{(N)}(d)$.*

Provided that we have a minimum parametrization there exists always a design d of size N, for $N \geq p$, such that $\mathbf{I}^{(N)}(d)$ is regular and hence every linear aspect ψ is identifiable under this design d.

If replications of the chosen level combinations occur a design d of size N may be alternatively represented by its J different level combinations $t^{(1)}, ..., t^{(J)}$ and the associated numbers $N_1, ..., N_J$ of replications, $\sum_{j=1}^{J} N_j = N$,

$$d \sim \begin{pmatrix} t^{(1)} & \cdots & t^{(J)} \\ N_1 & \cdots & N_J \end{pmatrix} .$$

The information matrix of d can, now, be written as $\mathbf{I}^{(N)}(d) = \sum_{j=1}^{J} w_j a(t^{(j)})a(t^{(j)})'$, where $w_j = \frac{N_j}{N}$ is the proportion of observations at the level combination $t^{(j)}$. More generally, we consider the normed measure $\delta_d = \sum_{j=1}^{J} w_j\epsilon_{t^{(j)}}$ associated with the design d where ϵ_t denotes the one-point measure at t. Then we may write the information matrix $\mathbf{I}^{(N)}(d) = \int a(t)a(t)'\delta_d(dt) = \int aa'd\delta_d$ as an integral with respect to δ_d.

It has been already mentioned by KIEFER and WOLFOWITZ (1959) that the set T^N of designs of size N is too sparse to be treated by the usual optimization techniques. In a first step we relax the assumption that Nw_j is integer and replace the fraction $\frac{N_j}{N}$ by arbitrary weights $w_j \geq 0$ with $\sum_{j=1}^J w_j = 1$.

δ is a discrete design *on T if $\delta = \sum_{j=1}^J w_j \epsilon_{t^{(j)}}$ for some $t^{(j)} \in T$ and weights $w_j \geq 0$, $\sum_{j=1}^J w_j = 1$.*

The set of discrete designs is convex and, hence, can be treated more easily. As this set is not closed we may consider the set of all probability measures on the design region T if it is complimented with a suitable σ-field.

Definition 1.1
δ is a (generalized) design *on T if δ is a probability measure on T. The set of all designs on T is denoted by Δ.*
$\mathbf{I}(\delta) = \int aa'd\delta$ is the information matrix *of the design δ.*

For every design d of size N the information matrices $\mathbf{I}^{(N)}(d)$ and $\mathbf{I}(\delta_d)$ for the associated (discrete) design δ_d coincide. Also the concept of identifiability has to be generalized appropriately,

Definition 1.2
(i) *A linear aspect $\psi : I\!R^p \to I\!R^s$ defined by $\psi(\beta) = L_\psi\beta$ is* identifiable *under δ if there exists a matrix $M_\psi \in I\!R^{s\times p}$ such that $L_\psi = M_\psi \mathbf{I}(\delta)$.*
(ii) *The set of all designs δ for which ψ is identifiable is denoted by $\Delta(\psi)$.*

In particular, $\Delta(\beta)$ is the set of all designs δ for which the whole parameter vector β is identifiable, i. e. for which $\mathbf{I}(\delta)$ is regular. We mention that the sets $\Delta(\psi)$ are convex. As we are essentially interested in the covariance of the best linear unbiased estimator we also generalize the concept of covariance matrices,

Definition 1.3
$\mathbf{C}_\psi(\delta) = L_\psi \mathbf{I}(\delta)^- L'_\psi$ is the (generalized) covariance matrix *of the design $\delta \in \Delta(\psi)$ for the linear aspect ψ.*
$\mathbf{C}(\delta) = \mathbf{I}(\delta)^{-1}$ is the covariance matrix of the design $\delta \in \Delta(\beta)$.

For notational convenience we will write sometimes $\Delta(L_\psi\beta)$ and $\mathbf{C}_{L_\psi\beta}$ instead of $\Delta(\psi)$ and $\mathbf{C}_\psi$, respectively. We note that for designs d of size N the covariance matrices $Cov(\widehat{\psi}_d)$ and $\mathbf{C}_\psi(\delta_d)$ for the associated (discrete) design δ_d coincide up to the multiplicative constant $\frac{1}{N}\sigma^2$. Moreover, the covariance matrix $\mathbf{C}_\psi(\delta)$ is independent of the special choice of the generalized inverse of the information matrix $\mathbf{I}(\delta)$ as ψ is identifiable.

Originally, the extension to discrete designs (or even generalized designs) has been introduced to embed the optimization problem into a richer set which is convex and closed. However, because the splitting of the experimental situation into N units with equally dispersed random errors is rather artificial a more general idea can be discovered behind the concept of discrete designs (see KUROTSCHKA (1988) and WIERICH (1989b)). Assume that we may observe at certain experimental conditions $t^{(1)}, ..., t^{(J)}$ with precision proportional to $w_j > 0$, i. e. $Var(X_j(t^{(j)})) = \frac{\sigma^2}{w_j}$, $\sum_{j=1}^J w_j = 1$, then in this general linear model the best linear unbiased estimator is given by the weighted least squares estimator $\widehat{\beta}_\delta = (A'_\delta W_\delta A_\delta)^{-1} A'_\delta W_\delta X_\delta$ (see e. g. CHRISTENSEN (1987), p 23), where $X_\delta = (X_j(t^{(j)}))_{j=1,\ldots,J}$ is the vector of observations, $A_\delta = (a_i(t^{(j)}))_{j=1,\ldots,J}^{i=1,\ldots,p}$ is the design matrix associated with the supporting points of the design $\delta = \sum_{j=1}^{J} w_j \epsilon_{t^{(j)}}$, and the diagonal matrix $W_\delta = \mathrm{diag}((w_j)_{j=1,\ldots,J})$ is the precision matrix for the observations. We notice that $\mathbf{I}(\delta) = A'_\delta W_\delta A_\delta$ and that $\delta \in \Delta(\psi)$ if and only if the linear aspect ψ is identifiable in the general linear model described above. In this model $\widehat{\psi}_\delta = L_\psi (A'_\delta W_\delta A_\delta)^- A'_\delta W_\delta X_\delta$ is the best linear unbiased estimator of ψ and $\frac{1}{\sigma^2} Cov(\widehat{\psi}_\delta) = L_\psi (A'_\delta W_\delta A_\delta)^- L'_\psi = \mathbf{C}_\psi(\delta)$. Hence, a discrete design $\delta = \sum_{j=1}^{J} w_j \epsilon_{t^{(j)}}$ has to be understood as an instruction which demands that J observations are to be made at the design points $t^{(j)}$ with precision proportional to w_j. This clarifies the definition of identifiability of ψ under δ because, now, ψ is identifiable if and only if ψ is a linear function of $A_\delta \beta$ which means that ψ has to be obtained from the values $\mu(t^{(j)}) = a(t^{(j)})'\beta$ of the response function evaluated at the supporting points $t^{(j)}$ of the design δ.

To avoid unnecessary technicalities we assume that the design region T is compact (or, at least, relatively compact) which is no real restriction because in practical applications either a finite number of different treatments are available for each factor or the range of the factors varying on a continuum is bounded by some experimental constraints. Furthermore, we only consider regression function a which are continuous on T. (Note that every function on a finite support is continuous.) Under these conditions even the subset of those information matrices $\mathbf{I}(\delta)$ associated with discrete designs is compact (see PÁZMAN (1986), p 60) and by CARATHÉODORY's theorem (see SILVEY (1980), pp 16, 72) we obtain an upper bound for the number of supporting points. Hence, the above conditions guarantee the existence of (discrete) optimum designs for the criteria under consideration. However, these conditions are not necessary for most of the following results to hold.

2 A Review on Optimum Design Theory

In this section we present the definition of various optimality criteria (Subsection 2.1) and collect the corresponding equivalence theorems which are useful tools for checking optimality (Subsection 2.2). We omit the proofs of these standard results since they have been presented extensively in the literature. In particular, we refer to the monographs by FEDOROV (1972), BANDEMER et al. (1977, 1980), ERMAKOV et al. (1983, 1987), SILVEY (1980), PÁZMAN (1986) and PUKELSHEIM (1993).

2.1 Optimality Criteria

In the most simple case where the unknown parameter β is one-dimensional it is common sense that a design δ^* is considered to be optimum if β is identifiable and the variance $\mathbf{C}(\delta)$ is minimized by δ^*. This design δ^* will produce the best linear unbiased estimator for β in this experimental situation and, additionally, in case of GAUSSian noise a uniformly most powerful test can be obtained compared to any other design δ. A natural extension to the multi-dimensional situation is given by the requirement that the variances $\mathbf{C}_{c'\beta}(\delta)$ for any one-dimensional linear combination $c'\beta$ of the unknown parameter β are minimized simultaneously. This is equivalent to the demand that $\mathbf{C}(\delta^*) \leq \mathbf{C}(\delta)$ for every $\delta \in \Delta(\beta)$, where the relation $\leq$ is meant in the positive-semidefinite sense, i. e. $C_1 \leq C_2$ if and only if $C_2 - C_1$ is positive-semidefinite. Actually, this uniform optimality $\mathbf{C}(\delta^*) \leq \mathbf{C}(\delta)$ cannot be established except in degenerate cases. Thus, we have to look for compromises.

In general, we will not be interested in all linear combinations $c'\beta$ but only in some features of them which might be summarized in a one-dimensional function of the covariance matrix $\mathbf{C}(\delta)$. Even, more generally, only some linear aspect ψ of β will be of interest and the optimality considerations will be based on the covariance matrix $\mathbf{C}_\psi(\delta)$ or a one-dimensional function of it.

Definition 2.1
A function $\Phi : \Delta \to I\!R \cup \infty$ *is a* criterion function *on the designs if* $\Phi(\delta_1) \leq \Phi(\delta_2)$ *for* $\mathbf{I}(\delta_1) \geq \mathbf{I}(\delta_2)$.
A design δ^* *is* Φ-optimum *if* $\Phi(\delta^*) \leq \Phi(\delta)$ *for every* $\delta \in \Delta$.

Note that the relation $\mathbf{I}(\delta_1) \geq \mathbf{I}(\delta_2)$ is equivalent to $\mathbf{C}(\delta_1) \leq \mathbf{C}(\delta_2)$ for $\delta_1, \delta_2 \in \Delta(\beta)$ and, moreover, implies $\mathbf{C}_\psi(\delta_1) \leq \mathbf{C}_\psi(\delta_2)$ for all $\delta_1, \delta_2 \in \Delta(\psi)$. This natural ordering of the information matrices is preserved by the criterion functions Φ which, with regard to that, can be written as a function of the information matrix, $\Phi(\delta) = \phi(\mathbf{I}(\delta))$. Next, we present a list of optimality criteria which will be considered throughout these notes and refer to the quoted literature for further interpretations. We start with the famous criterion of the *generalized variance* (determinant of the covariance matrix) the motivation of which can be derived from the power of the associated F-test and the size of confidence ellipsoids.

Definition 2.2
A design δ^* *is* D-optimum *if* $\det(\mathbf{I}(\delta^*)) \geq \det(\mathbf{I}(\delta))$ *for every* $\delta \in \Delta$.

The definition of D-optimality can equivalently be formulated as $\delta^* \in \Delta(\beta)$ and $\det(\mathbf{C}(\delta^*)) \leq \det(\mathbf{C}(\delta))$ for every $\delta \in \Delta(\beta)$. We note that compared to many other criteria the D-criterion has the advantage that it is not affected by reparametrizations of the model, i. e. if two parametrizations are given by $\mu(t) = a(t)'\beta = \widetilde{a}(t)'\widetilde{\beta}$, then there is a regular linear transformation matrix L such that $\widetilde{a} = La$ and the information matrices $\mathbf{I}(\delta)$ and $\widetilde{\mathbf{I}}(\delta)$ for the two parametrizations are linked by the equation $\widetilde{\mathbf{I}}(\delta) = L\mathbf{I}(\delta)L'$; hence $\det(\widetilde{\mathbf{I}}(\delta)) = \gamma \det(\mathbf{I}(\delta))$, where $\gamma = \det(L)^2$ is independent of δ, and the D-optimality carries over from one parametrization to the other.

A second very suggestive criterion is directly attached to the parameters and attempts to minimize the expected mean squared deviation of the estimates which is given by the trace of the covariance matrix $\mathbf{C}(\delta)$,

Definition 2.3
A design δ^* is A-optimum if $\mathrm{tr}(\mathbf{C}(\delta^*)) \leq \mathrm{tr}(\mathbf{C}(\delta))$ for every $\delta \in \Delta(\beta)$.

The A-criterion is sensitive to reparametrizations, but orthogonal ones will leave it unchanged because of $\mathrm{tr}(\widetilde{\mathbf{C}}(\delta)) = \mathrm{tr}(L'^{-1}\mathbf{C}(\delta)L^{-1}) = \mathrm{tr}(\mathbf{C}(\delta)(L'L)^{-1})$.

The next criterion is based on the minimax concept, the maximum variance of any linear combination of the parameter vector is to be minimized.

Definition 2.4
A design δ^* is E-optimum if $\lambda_{\min}(\mathbf{I}(\delta^*)) \geq \lambda_{\min}(\mathbf{I}(\delta))$ for every $\delta \in \Delta$.

This definition is equivalent to $\delta^* \in \Delta(\beta)$ and $\lambda_{\max}(\mathbf{C}(\delta^*)) \leq \lambda_{\max}(\mathbf{C}(\delta))$ for every $\delta \in \Delta(\beta)$ which is the same as $\max_{c'c=1} \mathrm{C}_{c'\beta}(\delta^*) \leq \max_{c'c=1} \mathrm{C}_{c'\beta}(\delta)$. Here, $\lambda_{\min}$ and $\lambda_{\max}$ denote the smallest resp. largest eigenvalue of the corresponding matrix. As $\det(\mathbf{C}(\delta)) = \prod_{i=1}^{p} \lambda_i(\mathbf{C}(\delta))$ and $\mathrm{tr}(\mathbf{C}(\delta)) = \sum_{i=1}^{p} \lambda_i(\mathbf{C}(\delta))$ all the criteria considered so far depend on the covariance matrix $\mathbf{C}(\delta)$ through its eigenvalues $\lambda_1(\mathbf{C}(\delta)), ..., \lambda_p(\mathbf{C}(\delta))$. They can be subsumed within the class of Φ_q-criteria which measure the q-"norm" of the vector of eigenvalues.

Definition 2.5
A design δ^* is Φ_q-optimum, if $\sum_{i=1}^{p} \lambda_i(\mathbf{C}(\delta^*))^q \leq \sum_{i=1}^{p} \lambda_i(\mathbf{C}(\delta))^q$ for every $\delta \in \Delta(\beta)$, $0 < q < \infty$.

For $q = 1$ we recover the A-criterion and the D- and E-criteria can be considered as limiting cases $q = 0$ and $q = \infty$ respectively (see e. g. KIEFER (1974) or PÁZMAN (1986), pp 94).

We may restrict the minimax approach as considered in the E-criterion to the design region T. In that case we are interested in minimizing the maximum variance of the pointwise prediction of the response function within the design region.

Definition 2.6
A design δ^* is G-optimum if $\max_{t \in T} \mathrm{C}_{a(t)'\beta}(\delta^*) \leq \max_{t \in T} \mathrm{C}_{a(t)'\beta}(\delta)$ for every $\delta \in \Delta(\beta)$.

This very suggestive criterion has already been used by SMITH (1918) in her pioneering paper on optimum design for polynomial regression. We may replace the maximum in Definition 2.6 by a weighted average (see e. g. FEDOROV (1972), p 142). For that purpose let ξ be a finite (weighting) measure on the design region T.

Definition 2.7

A design δ^* *is* Q-optimum *with respect to* ξ *if* $\int \mathbf{C}_{a(t)'\beta}(\delta^*)\,\xi(dt) \leq \int \mathbf{C}_{a(t)'\beta}(\delta)\,\xi(dt)$ *for every* $\delta \in \Delta_0$.

Mainly a Q-criterion is used when the design region T is a convex subset of $\mathbb{R}^K$ and ξ is some measure which is equivalent to the LEBESGUE measure $\lambda|_T$ restricted to T. Then the set Δ_0 of competing designs can be chosen as $\Delta(\beta)$ and the definition of Q-optimality is equivalent to $\int a(t)'\mathbf{C}(\delta^*)a(t)\,\xi(dt) \leq \int a(t)'\mathbf{C}(\delta)a(t)\,\xi(dt)$. Note that, in case ξ is concentrated on a smaller support, only the identifiability of the linear aspects $a(t)'\beta$ is required for t in the support of ξ. Obviously, both the G- and Q-criterion are not affected by any reparametrization of the model.

In many situations we are not interested in the whole parameter vector but merely in a linear aspect ψ. Hence the optimality criteria should depend on the covariance matrix $\mathbf{C}_\psi(\delta)$ for the linear aspect ψ rather than on $\mathbf{C}(\delta)$ and all designs $\delta \in \Delta(\psi)$ such that ψ is identifiable should be admitted as competitors.

Definition 2.8

(i) δ^* *is* D_ψ-optimum *(D-optimum for* ψ*) if* $\det(\mathbf{C}_\psi(\delta^*)) \leq \det(\mathbf{C}_\psi(\delta))$ *for every* $\delta \in \Delta(\psi)$.

(ii) δ^* *is* A_ψ-optimum *(A-optimum for* ψ*) if* $\mathrm{tr}(\mathbf{C}_\psi(\delta^*)) \leq \mathrm{tr}(\mathbf{C}_\psi(\delta))$ *for every* $\delta \in \Delta(\psi)$.

(iii) δ^* *is* E_ψ-optimum *(E-optimum for* ψ*) if* $\lambda_{\max}(\mathbf{C}_\psi(\delta^*)) \leq \lambda_{\max}(\mathbf{C}_\psi(\delta))$ *for every* $\delta \in \Delta(\psi)$.

(iv) δ^* *is* Φ_q-optimum *for* $\psi : \mathbb{R}^p \to \mathbb{R}^s$ *if* $\sum_{i=1}^s \lambda_i(\mathbf{C}_\psi(\delta^*))^q \leq \sum_{i=1}^s \lambda_i(\mathbf{C}_\psi(\delta))^q$ *for every* $\delta \in \Delta(\psi)$, $0 < q < \infty$.

In general, we will say that a design is Φ-optimum for ψ if it minimizes $\Phi(\delta)$ and Φ is a criterion function depending on $\mathbf{C}_\psi(\delta)$, $\Phi(\delta) = \varphi(\mathbf{C}_\psi(\delta))$ for $\delta \in \Delta(\psi)$ and $\Phi(\delta) = \infty$ otherwise. Again, the Φ_q-criteria for ψ include the A_ψ-criterion ($q = 1$) and the D_ψ- and E_ψ-criterion as limiting cases ($q = 0$ resp. $q = \infty$). We write for short $\Phi_{q;\psi}(\delta)$ for the numerical value of the Φ_q-criterion for ψ obtained by the design δ, e.g. $\Phi_{q;\psi}(\delta) = \sum_{i=1}^s \lambda_i(\mathbf{C}_\psi(\delta))^q$ for $\delta \in \Delta(\psi)$ and $0 < q < \infty$, and Φ_q for the whole parameter vector β. Note that the D_ψ-criterion makes only sense if the linear aspect ψ is *minimum*, i.e. if the corresponding matrix L_ψ has full row rank, $\mathrm{rank}(L_\psi) = s \leq p$. In case $s = p$ the D- and D_ψ-criteria are equivalent.

The Q-criterion can be recognized as a special case of the A_ψ-criterion for a particular linear aspect ψ. For this we notice that $\int \mathbf{C}_{a(t)'\beta}(\delta)\,\xi(dt) = \int a(t)'\mathbf{I}(\delta)^- a(t)\,\xi(dt) = \mathrm{tr}(\int aa'\,d\xi\,\mathbf{I}(\delta)^-)$ for every $\delta \in \Delta_0$. Now, $\int aa'\,d\xi$ is positive-semidefinite and can hence be decomposed into a matrix L_ψ and its transpose according to $\int aa'\,d\xi = L_\psi' L_\psi$. Thus, we get $\int \mathbf{C}_{a(t)'\beta}(\delta)\,\xi(dt) = \mathrm{tr}(L_\psi \mathbf{I}(\delta)^- L_\psi') = \mathrm{tr}(\mathbf{C}_\psi(\delta))$ and $\Delta_0 = \Delta_\psi$, where the linear aspect ψ is defined by $\psi(\beta) = L_\psi \beta$.

If the linear aspect ψ is one-dimensional, i.e. $\psi(\beta) = c'\beta$ for some $c \in \mathbb{R}^p$, then $\mathbf{C}_\psi(\delta) = \mathbf{C}_{c'\beta}(\delta) \in \mathbb{R}$ for $\delta \in \Delta(\psi)$ and all optimality criteria coincide.

Definition 2.9 δ^* *is* c-optimum *if* $\mathbf{C}_{c'\beta}(\delta^*) \leq \mathbf{C}_{c'\beta}(\delta)$ *for every* $\delta \in \Delta(c'\beta)$.

In particular, if β_i is one component of the parameter vector β we will call a design β_i-optimum if it minimizes the variance $\mathbf{C}_{\beta_i}(\delta)$ for β_i.

2.2 Equivalence Theorems

Already HOEL (1958) noticed that the D- and the G-optimum designs coincide in the model of a one-dimensional polynomial regression, and KIEFER and WOLFOWITZ (1960) proved that this is true for every linear model.

Theorem 2.1

δ^* is D-optimum if and only if $a(t)'\mathbf{C}(\delta^*)a(t) \leq p$ for every $t \in T$.

The equivalence of the D- and G-criterion now follows from $\mathbf{C}_{a(t)'\beta}(\delta) = a(t)'\mathbf{C}(\delta)a(t)$ and the fact that $\int a'\mathbf{C}(\delta)a\,d\delta = \mathrm{tr}(\int aa'\,d\delta\,\mathbf{C}(\delta)) = p$ and, hence, $\max_{t\in T} a(t)'\mathbf{C}(\delta)a(t) \geq p$ for every $\delta \in \Delta(\beta)$. The KIEFER-WOLFOWITZ equivalence theorem was complemented by an equivalence theorem for the A-optimality by FEDOROV (1971).

Theorem 2.2

δ^* is A-optimum if and only if $a(t)'\mathbf{C}(\delta^*)\mathbf{C}(\delta^*)a(t) \leq \mathrm{tr}(\mathbf{C}(\delta^*))$ for every $t \in T$.

Both these results can now be considered as applications of a general minimax theorem. To this end assume that we have an extended real valued function $\Phi : \Delta \to \mathbb{R} \cup \{\infty\}$ which is convex on the convex set Δ of designs on the design region T. Assume further that the directional derivatives $F_\Phi(\delta^*, \delta) = \lim_{\alpha\downarrow 0} \frac{1}{\alpha}(\Phi((1-\alpha)\delta^* + \alpha\delta) - \Phi(\delta^*))$ of Φ at δ^* in the direction of δ exist. Then the following general equivalence theorem due to WHITTLE (1973) holds,

Theorem 2.3

(i) δ^* minimizes $\Phi(\delta)$ if and only if $F_\Phi(\delta^*, \delta) \geq 0$ for every $\delta \in \Delta$.

(ii) If $F_\Phi(\delta^*, \cdot)$ is linear, i. e. $F_\Phi(\delta^*, \delta) = \int F_\Phi(\delta^*, \epsilon_t)\,\delta(dt)$ for every $\delta \in \Delta$, then δ^* minimizes $\Phi(\delta)$ if and only if $F_\Phi(\delta^*, \epsilon_t) \geq 0$ for every $t \in T$.

For more details see e. g. BANDEMER et al. (1977) or SILVEY (1980). Now, the Φ_q-criteria as introduced in Subsection 2.1, $0 \leq q \leq \infty$, including the A-, D- and E-criterion, are convex, at least, after some monotonic transformations of the criterion functions (see e. g. KIEFER (1974) or PÁZMAN (1986) for integer q). For example, Φ defined by $\Phi(\delta) = \ln\det(\mathbf{C}(\delta))$ for $\delta \in \Delta(\beta)$, $\Phi(\delta) = \infty$ otherwise, is a convex version of the criterion function for the D-criterion. Furthermore, the Φ_q-criteria, $0 \leq q < \infty$, are differentiable in the sense that the directional derivatives exist and are linear for every $\delta^* \in \Delta(\beta)$ (see e. g. KIEFER (1974)). With the E-criterion ($q = \infty$) problems arise, if the largest eigenvalue of $\mathbf{C}(\delta^*)$ is not simple (see KIEFER (1974)). These considerations remain true if the whole parameter vector is replaced by some linear aspect ψ and the optimum design results in a regular information matrix, $\delta^* \in \Delta(\beta)$.

Moreover, the criterion functions may be expressed in terms of a function ϕ of the information matrix $\mathbf{I}(\delta)$. Hence, if ϕ is an extended real valued function on the set $\mathbb{R}^{p\times p}_{(+)}$ of positive-semidefinite matrices which is convex and which is differentiable on the subset of positive-definite matrices with derivative ∇_ϕ, then the criterion function Φ defined by $\Phi(\delta) = \phi(\mathbf{I}(\delta))$ satisfies the conditions of Theorem 2.3. The according general equivalence theorem has been formulated by KIEFER (1974).

Theorem 2.4

Let $\Phi : \Delta \to \mathbb{R} \cup \{\infty\}$ be defined by $\Phi(\delta) = \phi(\mathbf{I}(\delta))$, where $\phi : \mathbb{R}^{p\times p}_{(+)} \to \mathbb{R} \cup \{\infty\}$. Assume that Φ is convex, $\delta^* \in \Delta(\beta)$, and that the derivative ∇_ϕ of ϕ at $\mathbf{I}(\delta^*)$ exist. Then δ^* minimizes $\Phi(\delta)$ if and only if $a(t)'\nabla_\phi(\mathbf{I}(\delta^*))a(t) \geq \mathrm{tr}(\mathbf{I}(\delta^*)\nabla_\phi(\mathbf{I}(\delta^*)))$ for every $t \in T$.

(For details see KIEFER (1974) or PÁZMAN (1986), p 117.) For example, these conditions are satisfied for $\phi = -\ln\det$ which corresponds to the D-criterion and Theorem 2.1 follows because of $\nabla_{\phi}(\mathbf{I}(\delta)) = -\mathbf{I}(\delta)^{-1}$ for $\delta \in \Delta(\beta)$. By the chain rule of differentiation for matrices (see e. g. GRAHAM (1981)) the result can be transferred to criterion functions Φ which are defined by a function φ of the covariance matrix $\mathbf{C}_{\psi}(\delta)$ for some linear aspect ψ (KIEFER (1974)).

Theorem 2.5

Let $\Phi : \Delta \to I\!R \cup \{\infty\}$ be defined by $\Phi(\delta) = \varphi(\mathbf{C}_{\psi}(\delta))$ for $\delta \in \Delta(\psi)$ and $\Phi(\delta) = \infty$ otherwise, where ψ is a minimum linear aspect with $\psi(\beta) = L_{\psi}\beta$. Assume that Φ is convex, $\delta^* \in \Delta(\beta)$, and that the derivative ∇_{φ} of φ at $\mathbf{C}_{\psi}(\delta^*)$ exist. Then δ^* minimizes $\Phi(\delta)$ if and only if $a(t)'\mathbf{C}(\delta^*)L_{\psi}'\nabla_{\varphi}(\mathbf{C}_{\psi}(\delta^*))L_{\psi}\mathbf{C}(\delta^*)a(t) \leq \mathrm{tr}(\mathbf{C}_{\psi}(\delta^*)\nabla_{\varphi}(\mathbf{C}_{\psi}(\delta^*)))$ for every $t \in T$.

For example $\varphi(\mathbf{C}(\delta)) = \mathrm{tr}(\mathbf{C}(\delta))$ which corresponds to the A-criterion satisfies the above conditions which establishes Theorem 2.2 because of $\nabla_{\varphi}(\mathbf{C}(\delta)) = \mathbf{E}_p$. Theorem 2.5 makes equivalence theorems available for a variety of criteria.

Corollary 2.6

Let $\delta^* \in \Delta(\beta)$ and let ψ be an s-dimensional minimum linear aspect, $\psi(\beta) = L_{\psi}\beta$. Then δ^* is D_{ψ}-optimum if and only if $a(t)'\mathbf{C}(\delta^*)L_{\psi}'\mathbf{C}_{\psi}(\delta^*)^{-1}L_{\psi}\mathbf{C}(\delta^*)a(t) \leq s$ for every $t \in T$.

Corollary 2.7

Let $\delta^* \in \Delta(\beta)$ and let ψ be a linear aspect, $\psi(\beta) = L_{\psi}\beta$. Then δ^* is A_{ψ}-optimum if and only if $a(t)'\mathbf{C}(\delta^*)L_{\psi}'L_{\psi}\mathbf{C}(\delta^*)a(t) \leq \mathrm{tr}(\mathbf{C}_{\psi}(\delta^*))$ for every $t \in T$.

Corollary 2.8

Let $\delta^* \in \Delta(\beta)$. Then δ^* is Q-optimum with respect to ξ if and only if $a(t)'\mathbf{C}(\delta^*)\int aa'\,d\xi\,\mathbf{C}(\delta^*)a(t) \leq \int a'\mathbf{C}(\delta^*)a\,d\xi$ for every $t \in T$.

Corollary 2.9

Let $\delta^* \in \Delta(\beta)$. Then δ^* is c-optimum if and only if $(a(t)'\mathbf{C}(\delta^*)c)^2 \leq c'\mathbf{C}(\delta^*)c$ for every $t \in T$.

All these equivalence theorems require that $\delta^* \in \Delta(\beta)$, i. e. that $\mathbf{I}(\delta^*)$ is regular. However, if we are interested in some linear aspect ψ rather than in the whole parameter vector β itself, then there is no reason why the whole parameter vector should be identifiable under the optimum design δ^*. To overcome that problem SILVEY (1978) and PUKELSHEIM and TITTERINGTON (1983) presented an equivalence theorem for possibly singular information matrices $\mathbf{I}(\delta^*)$. This theorem can be formulated in analogy to the general equivalence theorem (Theorem 2.5) by replacing the covariance matrix $\mathbf{C}(\delta^*)$ there by a special non-singular generalized inverse of the information matrix $\mathbf{I}(\delta^*)$.

Theorem 2.10

Let $\Phi : \Delta \to I\!R \cup \{\infty\}$ be defined by $\Phi(\delta) = \varphi(\mathbf{C}_{\psi}(\delta))$ for $\delta \in \Delta(\psi)$ and $\Phi(\delta) = \infty$ otherwise, where ψ is an s-dimensional minimum linear aspect with $\psi(\beta) = L_{\psi}\beta$. Assume that Φ is convex, $\delta^* \in \Delta(\beta)$, and that the derivative ∇_{φ} of φ at $\mathbf{C}_{\psi}(\delta^*)$ exists. Then

δ^* minimizes $\Phi(\delta)$ if and only if there exists a matrix H, such that $\mathrm{rank}(H) = p - s$, *$\mathbf{I}(\delta^*) + H'H$ is regular, and*

$$a(t)'(\mathbf{I}(\delta^*) + H'H)^{-1}L'_\psi\nabla_\varphi(\mathbf{C}_\psi(\delta^*))L_\psi(\mathbf{I}(\delta^*) + H'H)^{-1}a(t) \leq \mathrm{tr}(\mathbf{C}_\psi(\delta^*)\nabla_\varphi(\mathbf{C}_\psi(\delta^*)))$$

for every $t \in T$.

Further equivalence theorems in the singular case have been obtained e. g. by PÁZMAN (1980), NÄTHER and REINSCH (1981), and GAFFKE (1987). Conclusions similar to Corollaries 2.6 to 2.9 can be derived from Theorem 2.10 in a straightforward manner. For example, δ^* is D_ψ-optimum for an s-dimensional linear aspect ψ if there exists a matrix H such that $\mathrm{rank}(H) = p - s$, $\mathbf{I}(\delta^*) + H'H$ is regular, and $a(t)'(\mathbf{I}(\delta^*) + H'H)^{-1}L'_\psi\mathbf{C}_\psi(\delta^*)^{-1}L_\psi(\mathbf{I}(\delta^*) + H'H)^{-1}a(t) \leq s$ for every $t \in T$. Using that we can rather easily obtain the well-known helpful result that a D_ψ-optimum design with a minimum support, i. e. which is concentrated on s design points, assigns equal weights $\frac{1}{s}$ to its supporting points,

Lemma 2.11

If $\delta^* = \sum_{n=1}^{s} w_n\epsilon_{t^{(n)}}$ and if δ^* is D_ψ-optimum for an s-dimensional minimum linear aspect ψ, then $w_n = \frac{1}{s}$, for $n = 1, ..., s$.

Proof

We are going to show that for every D_ψ-optimum design $\delta^* = \sum_{n=1}^{N} w_n\epsilon_{t^{(n)}}$ the (positive) weights are bounded by $\frac{1}{s}$. In case of a minimum support ($N = s$) this proves the lemma.

First we notice that $\int a(t)'\mathbf{I}(\delta^*)^{-}L'_\psi\mathbf{C}_\psi(\delta^*)^{-1}L_\psi\mathbf{I}(\delta^*)^{-}a(t)\,\delta^*(dt) = s$ for any generalized inverse $\mathbf{I}(\delta^*)^{-}$ of $\mathbf{I}(\delta^*)$. In view of the above equivalence theorem there is a regular generalized inverse $\mathbf{I}(\delta^*)^{(-)}$ for which the integrand is bounded by s, hence $a(t^{(n)})'\mathbf{I}(\delta^*)^{(-)}L'_\psi\mathbf{C}_\psi(\delta^*)^{-1}L_\psi\mathbf{I}(\delta^*)^{(-)}a(t^{(n)}) = s$.

On the other hand we observe that $\mathbf{I}(\delta^*) \geq w_n a(t^{(n)})a(t^{(n)})'$. Thus by the identifiability of ψ we get $\mathbf{C}_\psi(\delta^*) = L_\psi\mathbf{I}(\delta^*)^{(-)}\mathbf{I}(\delta^*)\mathbf{I}(\delta^*)^{(-)}L'_\psi \geq w_n L_\psi\mathbf{I}(\delta^*)^{(-)}a(t^{(n)})a(t^{(n)})'\mathbf{I}(\delta^*)^{(-)}L'_\psi$. Now, letting $z_n = \mathbf{C}_\psi(\delta^*)^{-1}L_\psi\mathbf{I}(\delta^*)^{(-)}a(t^{(n)})$ and combining the above results we see that $s = z'_n\mathbf{C}_\psi(\delta^*)z_n \geq w_n(z'_n L_\psi\mathbf{I}(\delta^*)^{(-)}a(t^{(n)}))^2 = w_n s^2$ and hence $w_n \leq \frac{1}{s}$. □

For D-optimality, i. e. $s = p$, this result has been observed by HOEL (1958) for the model of a one-dimensional polynomial regression and for general models by KIEFER and WOLWOWITZ (1959). The general case has been established by ATWOOD (1973).

For the E-criterion KIEFER (1974) proved an equivalence theorem in the differentiable case when the largest eigenvalue of $\mathbf{C}(\delta^*)$ is simple. For the general case an equivalence theorem has been obtained by MELAS (1982) which can be presented in the following concise form (see PUKELSHEIM (1993) or ERMAKOV and MELAS (1995)).

Theorem 2.12

δ^* is E-optimum if and only if there exists a positive-semidefinite $p \times p$ matrix M such that $\mathrm{tr}(M) = 1$ and $a(t)'Ma(t) \leq \lambda_{\max}(\mathbf{C}(\delta^*))^{-1}$ for every $t \in T$.

Although all these equivalence theorems are very useful for verifying the optimality of a prespecified design their practical applicability is restricted if we are in a situation where we have to search the potentially optimum design first. The following equivalence result on c-optimality due to ELFVING (1952) is an exception and provides us with a rule for constructing the optimum design.

Theorem 2.13

Let $\delta^* = \sum_{n=1}^N w_n \epsilon_{t(n)}$. δ^* is c-optimum if and only if there exists a $\gamma > 0$ such that γc lies on the boundary of the convex hull of $\{a(t); t \in T\} \cup \{-a(t); t \in T\}$ and $\gamma c = \sum_{n=1}^N z_n w_n a(t^{(n)})$ for some $z_n \in \{-1, 1\}$, $n = 1, ..., N$. Moreover, in this case $\mathrm{C}_{c'\beta}(\delta^*) = \gamma^{-2}$.

The convex hull of $\{a(t); t \in T\} \cup \{-a(t); t \in T\}$ is called the ELFVING set associated with the regression function a. For further details see PUKELSHEIM (1981) or PÁZMAN (1986, p 71). ELVFING's theorem is constructive in so far that for given c we can find γc on the boundary of the convex hull and γc can always be represented by a convex combination $\sum_{n=1}^N z_n w_n a(t^{(n)})$ of some $z_n a(t^{(n)})$ which gives the design points $t^{(1)}, ..., t^{(N)}$ and the corresponding weights $w_1, ..., w_N$. In case that a constant term is involved in the model the statement can be considerably simplified for c in the convex hull of $\{a(t); t \in T\}$,

Corollary 2.14

Let $\mu(t) = \beta_0 + f(t)'\beta_1$, $t \in T$, let $c = (1, c_1')' \in \mathbb{R}^p$ and let c_1 lie in the convex hull of $\{f(t); t \in T\}$. Then the design $\delta^* = \sum_{n=1}^N w_n \epsilon_{t(n)}$ is c-optimum if and only if $c_1 = \sum_{n=1}^N w_n f(t^{(n)})$. Moreover, in this case $\mathrm{C}_{c'\beta}(\delta^*) = 1$.

Example 2.1 *Linear regression (cf Example 1.2 (i)):*

In the model $\mu(t) = \beta_0 + \beta_1 t$ of linear regression on a general interval $T = [\tau_1, \tau_2]$ an optimum design is sought for the prediction at a fixed setting τ. In case of interpolation, i.e. $\tau \in T$, a natural choice for an optimum design is ϵ_τ which assigns all weight to the setting τ of interest. It is obvious that $\mathrm{C}_{a(\tau)'\beta}(\epsilon_\tau) = 1$ which cannot be improved upon in view of Corollary 2.14. However, the optimum design is not unique, because $\tau = \frac{\tau_2 - \tau}{\tau_2 - \tau_1}\tau_1 + \frac{\tau - \tau_1}{\tau_2 - \tau_1}\tau_2$ is a convex combination of the endpoints τ_1 and τ_2 of the interval. Hence, the design $\delta^* = \frac{\tau_2 - \tau}{\tau_2 - \tau_1}\epsilon_{\tau_1} + \frac{\tau - \tau_1}{\tau_2 - \tau_1}\epsilon_{\tau_2}$ which assigns the weights obtained from the convex combination to the endpoints is also optimum for the linear aspect $a(\tau)'\beta$ by Corollary 2.14.

In case of extrapolation, $\tau \notin T$, the Corollary 2.14 is no longer applicable and we have to use the general form of ELFVING's Theorem. In fact, the ELFVING set is the convex hull spanned by the four elements $(1, \tau_1)$, $(1, \tau_2)$, $(-1, -\tau_1)$ and $(-1, -\tau_2)$ associated with the two endpoints of the interval. Hence, for every c there is a c-optimum design $\delta^* = w_1 \epsilon_{\tau_1} + w_2 \epsilon_{\tau_2}$ supported on those two points and the optimum weights are obtained by evaluating the conditions in Theorem 2.13, $w_1 = |\frac{\tau - \tau_2}{2\tau - (\tau_1 + \tau_2)}|$ and $w_2 = |\frac{\tau - \tau_1}{2\tau - (\tau_1 + \tau_2)}|$ for $c = (1, \tau)$, $\tau \notin T$, and the corresponding variance becomes $\mathrm{C}_{a(\tau)'\beta} = \frac{(2\tau - (\tau_1 + \tau_2))^2}{(\tau_2 - \tau_1)^2}$. □

3 Reduction Principles

In general, the set of all competing designs is rather large. Therefore, it is necessary to use tools which simplify the characterization of optimum designs and which permit to search for an optimum design in a substantially smaller subclass. The first part of this section is devoted to tools which deal with reductions to subsystems of regression functions and, hence, to subsystems of parameters. In the second part a further inherent structure is assumed in the underlying model which allows for reduction by invariance with respect to suitable transformations on the design region.

3.1 Orthogonalization and Refinement

In many situations there is a natural way how to split the set of regression functions $a = (a_1, ..., a_p)'$ into two parts $f_0 = (a_1, ..., a_{p_0})'$ and $f_1 = (a_{p_0+1}, ..., a_p)'$ and, hence, how to split the parameter vector β, if e. g. $f_0 = \mathbf{1}$ is a constant function, i. e. β_0 is a constant or intercept term in the model, or f_0 and f_1 are associated with the effects of different factors of influence as exhibited in Section 5. Conversely, β_0 may contain all parameters of interest and β_1 is regarded as a vector of noise parameters which induces a separation of the associated regression functions. In any of these cases we can rewrite the linear model $\mu(t) = a(t)'\beta$ as

$$\mu(t) = f_0(t)'\beta_0 + f_1(t)'\beta_1 \ , \tag{3.1}$$

$t \in T$, with $a = \binom{f_0}{f_1}$ and $\beta = \binom{\beta_0}{\beta_1}$. Before we introduce the general concepts of reduction to subsystems of regression functions we start with a preparatory result which is helpful in this context,

Lemma 3.1

For every design δ and every pair of regression functions f_0 and f_1 the identity $\int f_0 f_1' \, d\delta = \int f_0 f_0' \, d\delta (\int f_0 f_0' \, d\delta)^- \int f_0 f_1' \, d\delta$ holds, where $(\int f_0 f_0' \, d\delta)^-$ may be any generalized inverse of $\int f_0 f_0' \, d\delta$.

Proof

The information matrix $\mathbf{I}(\delta) = \int \binom{f_0}{f_1}(f_0', f_1') \, d\delta$ is positive-semidefinite and symmetric, and the result follows immediately from Lemma A.1. □

The roles of f_0 and f_1 may be interchanged, $\int f_1 f_0' \, d\delta = \int f_1 f_1' \, d\delta (\int f_1 f_1' \, d\delta)^- \int f_1 f_0' \, d\delta$, and, in particular, with $f_0 = \mathbf{1}$, this yields $\int f_1 \, d\delta = \int f_1 f_1' \, d\delta (\int f_1 f_1' \, d\delta)^- \int f_1 \, d\delta$.

As a first concept we introduce the *orthogonalization* with respect to a given design which dates back, at least, to the early works of KIEFER (1959) and KIEFER and WOLFOWITZ (1959) in which the orthogonalization was performed on a given (minimum) support. This technique has been applied successfully in both analysis of variance models with purely qualitative factors and in regression models on various design regions.

We will start with the elementary situation of a linear model $\mu(t) = \beta_0 + f(t)'\beta_1$, $t \in T$, with constant term in which f_0 is the constant function equal to one, $f_0 = \mathbf{1}$, and $f_1 = f$. The *centered version* $\widetilde{f}$ of the regression function f with respect to the design δ is defined by

$$\widetilde{f}(t) = f(t) - \int f \, d\delta \ . \tag{3.2}$$

By construction the centered version $\widetilde{f}$ is *orthogonal* to the constant function with respect to the design δ, i.e. $\int \widetilde{f}\, d\delta = \mathbf{0}$. The response function μ can be written as $\mu(t) = \widetilde{\beta}_0 + \widetilde{f}(t)'\widetilde{\beta}_1$, $t \in T$, where $\widetilde{\beta}_1 = \beta_1$ and $\widetilde{\beta}_0 = \beta_0 + \int f'\, d\delta\, \beta_1$. Hence, the centering causes a reparametrization of the original model which leaves β_1 unchanged. Then

$$\widetilde{\mathbf{I}}(\delta) = \begin{pmatrix} 1 & \mathbf{0} \\ \mathbf{0} & \int \widetilde{f}\widetilde{f}'\, d\delta \end{pmatrix}$$

is the information matrix with respect to this new parametrization (depending on δ).

Lemma 3.2

Let $\widetilde{f}$ be defined by (3.2) and let $J = \int ff'\, d\delta - \int f\, d\delta \int f'\, d\delta$.

(i) $\det(\mathbf{I}(\delta)) = \det(\widetilde{\mathbf{I}}(\delta)) = \det(J)$.

(ii) *The matrix*

$$\mathbf{I}(\delta)^- = \begin{pmatrix} 1 + \int f'\, d\delta\, J^- \int f\, d\delta & -\int f'\, d\delta\, J^- \\ -J^- \int f\, d\delta & J^- \end{pmatrix}$$

is a generalized inverse of the information matrix $\mathbf{I}(\delta)$, where J^- is an arbitrary generalized inverse of J.

(iii) $\delta \in \Delta(\beta)$ *if and only if* $\delta \in \Delta(\beta_1)$.

Proof

The matrix

$$\widetilde{L} = \begin{pmatrix} 1 & \mathbf{0} \\ -\int f\, d\delta & \mathbf{E}_{p-1} \end{pmatrix}.$$

is the transformation matrix for the reparametrization $\widetilde{a} = \widetilde{L}a$. Thus, the information matrices in the original and in the transformaed model are related by $\widetilde{\mathbf{I}}(\delta) = \widetilde{L}\mathbf{I}(\delta)\widetilde{L}'$. The transformation matrix $\widetilde{L}$ is lower triangular with all its diagonal elements equal to 1. Hence, $\widetilde{L}$ is regular and $\det(\widetilde{L}) = 1$. Thus, $\det(\widetilde{\mathbf{I}}(\delta)) = \det(\widetilde{L})^2 \det(\mathbf{I}(\delta)) = \det(\mathbf{I}(\delta))$ and the second equality follows from the block diagonal structure of $\widetilde{\mathbf{I}}(\delta)$ and $\int \widetilde{f}\widetilde{f}'\, d\delta = J$.

A generalized inverse

$$\widetilde{\mathbf{I}}(\delta)^- = \begin{pmatrix} 1 & \mathbf{0} \\ \mathbf{0} & J^- \end{pmatrix}$$

of the transformed information matrix $\widetilde{\mathbf{I}}(\delta)$ can be obtained by blockwise inversion. The proposed representation $\mathbf{I}(\delta)^- = \widetilde{L}'\widetilde{\mathbf{I}}(\delta)^-\widetilde{L}$ follows by verification of $\mathbf{I}(\delta)\widetilde{L}'\widetilde{\mathbf{I}}(\delta)^-\widetilde{L}\mathbf{I}(\delta) = \widetilde{L}^{-1}\widetilde{\mathbf{I}}(\delta)\widetilde{\mathbf{I}}(\delta)^-\widetilde{\mathbf{I}}(\delta)\widetilde{L}'^{-1} = \widetilde{L}^{-1}\widetilde{\mathbf{I}}(\delta)\widetilde{L}'^{-1} = \mathbf{I}(\delta)$.

If $\delta \in \Delta(\beta_1)$ this is preserved under the reparametrization induced by the centering because β_1 is left unchanged. Moreover, $\mathbf{C}_{\beta_1}(\delta) = J^{-1}$ by (ii) and, hence, both the transformed and the original information matrix are regular which proves $\delta \in \Delta(\beta)$. The converse implication is obvious. □

As a straightforward consequence we obtain the well-known useful result that the D-criterion for the linear aspect in which the parameter related to the constant term is omitted and the D-criterion for the whole parameter vector coincide (see e.g. STUDDEN (1980) or SCHWABE and WIERICH (1995)),

Theorem 3.3

In the model $\mu(t) = \beta_0 + f(t)'\beta_1$, $t \in T$, with a constant term the design δ^* is D-optimum if and only if δ^* is D_{β_1}-optimum.

Proof

In view of Lemma 3.2 we have $\mathbf{C}_{\beta_1}(\delta) = (\int ff'\,d\delta - \int f\,d\delta \int f'\,d\delta)^{-1}$ and, hence, $\det(\mathbf{I}(\delta)) = \det(\mathbf{C}_{\beta_1}(\delta))^{-1}$ which proves the equivalence of the criteria. □

For general linear models $\mu(t) = f_0(t)'\beta_0 + f_1(t)'\beta_1$, $t \in T$, results can be obtained similar to Lemma 3.2. In this situation $\widetilde{f}_1$ defined by

$$\widetilde{f}_1(t) = f_1(t) - \int f_1 f_0'\,d\delta (\int f_0 f_0'\,d\delta)^- f_0(t) \tag{3.3}$$

is a *centered version* of the regression function f_1 with respect to f_0 and the design δ, where $(\int f_0 f_0'\,d\delta)^-$ is an arbitrary generalized inverse of the partial information matrix $\mathbf{I}_0(\delta) = \int f_0 f_0'\,d\delta$ in the *submodel*

$$\mu_0(t) = f_0(t)'\beta_0\ , \tag{3.4}$$

$t \in T$. The centered version $\widetilde{f}_1$ is *orthogonal* to f_0 with respect to the design δ, i.e. $\int \widetilde{f}_1 f_0'\,d\delta = \mathbf{0}$, in view of Lemma 3.1. The response function μ can be written as $\mu(t) = f_0'\widetilde{\beta}_0 + \widetilde{f}_1(t)'\widetilde{\beta}_1$, $t \in T$, where $\widetilde{\beta}_1 = \beta_1$ and $\widetilde{\beta}_0 = \beta_0 + (\int f_0 f_0'\,d\delta)^- \int f_0 f_1'\,d\delta\,\beta_1$. Again, the centering causes a reparametrization of the original model which leaves β_1 unchanged and

$$\widetilde{\mathbf{I}}(\delta) = \begin{pmatrix} \int f_0 f_0'\,d\delta & \mathbf{0} \\ \mathbf{0} & \int \widetilde{f}_1 \widetilde{f}_1'\,d\delta \end{pmatrix}$$

is the information matrix with respect to this new parametrization (depending on δ).

Lemma 3.4

Let $\widetilde{f}_1$ be defined by (3.3) and $J = \int f_1 f_1'\,d\delta - \int f_1 f_0'\,d\delta (\int f_0 f_0'\,d\delta)^- \int f_0 f_1'\,d\delta$.

(i) $\det(\mathbf{I}(\delta)) = \det(\widetilde{\mathbf{I}}(\delta)) = \det(J)\det(\mathbf{I}_0(\delta))$.

(ii) *The matrix*

$$\mathbf{I}(\delta)^- = \begin{pmatrix} \mathbf{I}_0(\delta)^- + \int f_0 f_1'\,d\delta\, J^- \int f_1 f_0'\,d\delta & -\int f_0 f_1'\,d\delta\, J^- \\ -J^- \int f_1 f_0'\,d\delta & J^- \end{pmatrix}$$

is a generalized inverse of the information matrix $\mathbf{I}(\delta)$, where $\mathbf{I}_0(\delta)^-$ and J^- are arbitrary generalized inverses of $\mathbf{I}_0(\delta)$ and J, respectively.

(iii) *If $\delta \in \Delta(\beta_1)$ then $\mathbf{C}_{\beta_1}(\delta) = J^{-1}$.*

Proof

The proof follows along the same lines as indicated in Lemma 3.2. The transformation matrix

$$\widetilde{L} = \begin{pmatrix} \mathbf{E}_{p_0} & \mathbf{0} \\ -\int f_1 f_0'\,d\delta\, \mathbf{I}_0(\delta)^- & \mathbf{E}_{p-p_0} \end{pmatrix}$$

for the reparametrization $\widetilde{a} = \widetilde{L}a$ is, again, lower triangular with all its diagonal elements equal to 1. Hence, $\widetilde{L}$ is regular and $\det(\widetilde{\mathbf{I}}(\delta)) = \det(\widetilde{L})^2 \det(\mathbf{I}(\delta)) = \det(\mathbf{I}(\delta))$ The second equality in (i) follows from the block diagonal structure of $\widetilde{\mathbf{I}}(\delta)$ and $\int \widetilde{f}\widetilde{f}'\,d\delta = J$ in view of Lemma 3.1. The matrix

$$\widetilde{\mathbf{I}}(\delta)^- = \begin{pmatrix} \mathbf{I}_0(\delta) & \mathbf{0} \\ \mathbf{0} & J^- \end{pmatrix}$$

is a generalized inverse of the transformed information matrix $\widetilde{\mathbf{I}}(\delta)$ and the representation $\mathbf{I}(\delta)^- = \widetilde{L}'\widetilde{\mathbf{I}}(\delta)^-\widetilde{L}$ follows by verification of $\mathbf{I}(\delta)\widetilde{L}'\widetilde{\mathbf{I}}(\delta)^-\widetilde{L}\mathbf{I}(\delta) = \mathbf{I}(\delta)$ as before. Statement (iii) follows immediately from (ii). □

A second tool which is commonly used for the derivation of optimum designs is the concept of refinement considerations which decrease the complexity of the model. The main refinement argument can be summarized as follows: In a submodel where less regression functions are involved the performance of a design is, at least, as good as in the complete original model (Lemma 3.5). Hence, if a design δ^* is optimum in the submodel and its performance is the same for both the complete and the submodel, then this design δ^* is also optimum in the original model (Theorem 3.6; for further applications see e. g. MAGDA (1980) and KUNERT (1983)). A more detailed treatment of the relation between submodels and the complete model is given by BAKSALARY (1984).

Besides the complete model $\mu(t) = f_0(t)'\beta_0 + f_1(t)'\beta_1$, $t \in T$, we consider the submodel $\mu(t) = f_0(t)'\beta_0$, $t \in T$, of equation (3.4) in which the regression function $f_0 : T \to \mathbb{R}^{p_0}$, $p_0 < p$, is formed by a proper subset of the components $a_1, ..., a_p$ of the original regression function $a = \binom{f_0}{f_1}$. Similar to the information matrix $\mathbf{I}_0 = \int f_0 f_0'\,d\delta$ we define the covariance matrix of δ for a linear aspect ψ_0 by $\mathbf{C}_{0,\psi_0}(\delta) = L_{0,\psi_0}(\int f_0 f_0'\,d\delta)^- L_{0,\psi_0}'$ if ψ_0 is identifiable under δ in the submodel, $\psi_0(\beta_0) = L_{0,\psi_0}\beta_0$.

Lemma 3.5

Let ψ be a linear aspect of $\beta = \binom{\beta_0}{\beta_1}$ in the complete model $\mu(t) = f_0(t)'\beta_0 + f_1(t)'\beta_1$, let $L_\psi = (L_0|L_1)$ be partitioned in the same way as β, i. e. $\psi(\beta) = L_0\beta_0 + L_1\beta_1$ and let ψ_0 be the associated linear aspect in the submodel $\mu_0(t) = f_0(t)'\beta_0$, $\psi_0(\beta_0) = L_0\beta_0$.

(i) *If ψ is identifiable under δ in the complete model, then ψ_0 is identifiable under δ in the submodel.*

(ii) $\mathbf{C}_{0,\psi_0}(\delta) \leq \mathbf{C}_\psi(\delta)$ *for every $\delta \in \Delta(\psi)$.*

Proof

(i) Because ψ is identifiable there exists a matrix M_ψ satisfying $L_\psi = M_\psi\mathbf{I}(\delta)$. This matrix $M_\psi = (M_0|M_1)$ can be partitioned similarly to L_ψ. With this notation $L_0 = M_0\mathbf{I}_0(\delta) + M_1\int f_1 f_0'\,d\delta = (M_0 + M_1\int f_1 f_0'\,d\delta\mathbf{I}_0(\delta)^-)\mathbf{I}_0(\delta)$ by Lemma 3.1 and ψ_0 is identifiable in the submodel.

(ii) By the formula for a generalized inverse of a partitioned matrix (Lemma A.3) we get

$$\begin{aligned} L\mathbf{I}(\delta)^-L' &= L_0\mathbf{I}_0(\delta)^-L_0' + (L_1 - L_0\mathbf{I}_0(\delta)^-\textstyle\int f_0 f_1'\,d\delta) \\ &\quad \times (\textstyle\int f_1 f_1'\,d\delta - \int f_1 f_0'\,d\delta\mathbf{I}_0(\delta)^-\int f_0 f_1'\,d\delta)^-(L_1 - L_0\mathbf{I}_0(\delta)^-\int f_0 f_1'\,d\delta)' \\ &\geq L_0\mathbf{I}_0(\delta)^-L_0' \end{aligned}$$

which proves (ii). □

As can be seen from the proof equality is attained in Lemma 3.5 if (and only if) $L_1 = L_0\mathbf{I}_0(\delta)^-\int f_0 f_1'\,d\delta$.

Theorem 3.6

Let ψ be as in Lemma 3.5, let $\Phi(\delta) = \varphi(\mathbf{C}_\psi(\delta))$ and $\Phi^{(0)}(\delta) = \varphi(\mathbf{C}_{0,\psi_0}(\delta))$, for $\delta \in \Delta(\psi)$, $\Phi(\delta) = \infty$ otherwise. If the design δ^* minimizes $\Phi^{(0)}(\delta)$ and if $\Phi(\delta^*) = \Phi^{(0)}(\delta^*)$ then the design δ^* minimizes also $\Phi(\delta)$.

Proof

By Lemma 3.5 and the monotonicity of φ we have $\Phi(\delta) \geq \Phi^{(0)}(\delta) \geq \Phi^{(0)}(\delta^*) = \Phi(\delta^*)$ for every $\delta \in \Delta(\psi)$. □

Lemma 3.5 and Theorem 3.6 are of particular interest for linear aspects ψ which depend on the joint parameter β_0, $\psi(\beta) = \psi_0(\beta_0)$. For this case equality is attained, in particular, if f_0 and f_1 are *orthogonal* with respect to δ, $\int f_0 f_1' \, d\delta = \mathbf{0}$. The following results on D-, A- and E-optimality are straightforward consequences of Theorem 3.6,

Corollary 3.7

Let ψ be a linear aspect which depends on β_0, $\psi(\beta) = \psi_0(\beta_0)$.

(i) *If δ^* is D-optimum for ψ_0 in the submodel and if* $\det(\mathbf{C}_\psi(\delta^*)) = \det(\mathbf{C}_{0,\psi_0}(\delta^*))$ *then δ^* is D-optimum for ψ.*

(ii) *If δ^* is A-optimum for ψ_0 in the submodel and if* $\mathrm{tr}(\mathbf{C}_\psi(\delta^*)) = \mathrm{tr}(\mathbf{C}_{0,\psi_0}(\delta^*))$ *then δ^* is A-optimum for ψ.*

(iii) *If δ^* is E-optimum for ψ_0 in the submodel and if* $\lambda_{\max}(\mathbf{C}_\psi(\delta^*)) = \lambda_{\max}(\mathbf{C}_{0,\psi_0}(\delta^*))$ *then δ^* is E-optimum for ψ.*

Example 3.1 *Linear regression (cf Example 1.2 (i)):*

As a first application of the refinement argument we want to determine the D-optimum design in the linear regression model $\mu(t) = \beta_0 + \beta_1 t$ on the standardized design region $T = [-1, 1]$. In connection with that we consider the submodel $\mu_0(t) = \beta_1 t$. (Note that we have deliberately interchanged the roles of β_0 and β_1.) As the parameter β_1 of the submodel is one-dimensional all optimality criteria coincide and aim at maximizing the quantity $\int f^2 \, d\delta = \int t^2 \delta(dt)$ in the pressent setting. Hence, every design which is supported on the points -1 and 1 of maximum modulus of the regression function f, $f(t) = t$, is optimum in the submodel. Within this class of designs the regression function f is centered with respect to the symmetric design δ^* which assigns equal weights $\frac{1}{2}$ to both endpoints of the interval. Hence, the variances $\mathbf{C}_{\beta_1}(\delta^*) = 1 = \mathbf{C}_0(\delta^*)$ for β_1 coincide in both models, and δ^* is optimum for β_1 in the larger model in view of Corollary 3.7. Now, by Theorem 3.3 the design δ^* is D-optimum for the whole parameter vector in the (complete) model of linear regression.

For linear regression on an arbitrary interval $T = [\tau_1, \tau_2]$ we note that translations and scale transformations lead to linear transformations of the underlying linear regression model and, hence, the D-optimality is not affected. That means that the correspondingly transformed optimum design which assigns equal weights $\frac{1}{2}$ to each of the endpoints τ_1 and τ_2 of the interval is D-optimum on $T = [\tau_1, \tau_2]$. □

The following direct application of Lemma 3.5, where $\psi(\beta)$ may differ from $\psi_0(\beta_0)$, gives a lower bound for the variance in case of a one-dimensional linear aspect,

Lemma 3.8

Let $f_0 = 1$, i. e. $\mu(t) = \beta_0 + f_1(t)'\beta_1$, and let $\psi(\beta) = c'\beta = c_0 + c_1'\beta$, $c \in \mathbb{R}^p$. Then $\mathbf{C}_{c'\beta}(\delta) \geq c_0^2$.

We close this subsection with a result which relates the determinants in the complete and in the submodel for a certain class of linear aspects and is, hence, complementary to Lemma 3.4 (i),

Lemma 3.9

Let ψ be a linear aspect of β which contains the whole subsytem β_1, $\psi(\beta) = \binom{\psi_0(\beta_0)}{\beta_1}$. Then $\det \mathbf{C}_\psi(\delta) = \det \mathbf{C}_{0,\psi_0}(\delta) \det \mathbf{C}_{\beta_1}(\delta)$ *for every $\delta \in \Delta(\psi)$.*

Proof

By Lemma 3.5 the linear aspect ψ_0 is identifiable under δ in the submodel, $\psi_0(\beta_0) = L_{0,\psi_0}\beta_0$. We partition the covariance matrix

$$\mathbf{C}_\psi(\delta) = \begin{pmatrix} C_0 & C_{01} \\ C_{01}' & \mathbf{C}_{\beta_1}(\delta) \end{pmatrix}$$

according to the components ψ_0 and β_1 of ψ. The determinant of $\mathbf{C}_\psi(\delta)$ factorizes, $\det(\mathbf{C}_\psi(\delta)) = \det(C_0 - C_{01}\mathbf{C}_{\beta_1}(\delta)^{-1}C_{01}') \det(\mathbf{C}_{\beta_1}(\delta))$, by the corresponding formula for partitioned matrices (Lemma A.2). Further, we obtain $C_{01} = L_{0,\psi_0}\mathbf{I}_0(\delta)^- \int f_0 f_1' \, d\delta \, \mathbf{C}_{\beta_1}(\delta)$ and $C_0 = L_{0,\psi_0}(\mathbf{I}_0(\delta)^- + \mathbf{I}_0(\delta)^- \int f_0 f_1' \, d\delta \, \mathbf{C}_{\beta_1}(\delta) \int f_1 f_0' \, d\delta \, \mathbf{I}_0(\delta)^-)L_{0,\psi_0}'$ by the formula for generalized inverses of partitioned matrices (Lemma A.3) and, hence, $\mathbf{C}_{0,\psi_0}(\delta) = C_0 - C_{01}\mathbf{C}_{\beta_1}(\delta)^{-1}C_{01}'$ which completes the proof. □

3.2 Invariance

Considerations of invariance have been used for the reduction of statistical problems for long time (see e. g. LEHMANN (1959), chapter 6) and are still object and tool of scientific investigations (cf EATON (1989) and WIJSMAN (1990), besides others). Invariance considerations lead to essentially complete class theorems which allow a restriction to a substantially smaller subset of competing designs. (Note that the first essentially complete class results have been introduced to the theory of experimental design by EHRENFELD (1956).) For finding optimum designs invariance has been applied successfully by many authors including KIEFER (1959), KIEFER and WOLFOWITZ (1959), and ATWOOD (1969). For more recent treatments see PUKELSHEIM (1987, 1993), or GAFFKE and HEILIGERS (1995).

Here, we want to present a survey on the concept of invariance which will be kept as elementary as possible but also as general as necessary. In particular, we will confine to finite groups of transformations without further mentioning and refer the interested reader to the literature (cf HALMOS (1974), EATON (1989), or WIJSMAN (1990)) for the more general concept of compact groups which will be shortly illustrated by an example at the end of this subsection. Additionally, for polynomial regression models some majorization arguments are included (for a detailed treatment we refer to GIOVAGNOLI, PUKELSHEIM, and WYNN (1987)).

Throughout the general linear model $\mu(t) = a(t)'\beta$, $t \in T$, is considered and we start with a couple of definitions concerning the corresponding invariance concepts.

Definition 3.1

(i) *A one-to-one mapping $g : T \to T$ is called a* transformation *of the design region T.*

(ii) *A transformation g of T* induces a linear transformation *of the regression function $a : T \to \mathbb{R}^p$ if there exists a $p \times p$ matrix Q_g with $a(g(t)) = Q_g a(t)$ for every $t \in T$.*

Usually we will be concerned with a whole group G of transformations (e. g. permutations),

Definition 3.2

A group G of transformations of T induces linear transformations *of the regression function $a : T \to \mathbb{R}^p$ if every $g \in G$ induces a linear transformation of a.*

In the situation of Definition 3.2 we will say that the linear model is linearly transformed by G because every transformation $g \in G$ causes a reparametrization of the model $\mu(t) = a(t)'\beta = a(g(t))'\widetilde{\beta}^g$, where $\widetilde{\beta}^g = Q_g'^{-1}\beta$. With G also $\{Q_g;\, g \in G\}$ is a group because we have assumed that the components of the regression function a are linearly independent. In particular, Q_g is regular for every $g \in G$ and $Q_{g^{-1}} = Q_g^{-1}$, where g^{-1} is the inverse transformation to g. By $\#G$ we denote the number of transformations in the (finite) group G.

Next, we notice that for every design δ on the design region T the induced design δ^g, defined by $\delta^g(B) = \delta(g^{-1}(B))$ for any (measurable) subset B of T, is again a design on the design region T.

Definition 3.3

A design δ is invariant *with respect to G if $\delta^g = \delta$ for every $g \in G$.*

In case G induces linear transformations of a we obtain $\mathbf{I}(\delta^g) = \int a(t)a(t)'\delta^g(dt) = \int a(g(t))a(g(t))'\delta(dt) = Q_g\mathbf{I}(\delta)Q_g'$ for the information matrix of the induced design δ^g. Hence, we can weaken the requirements of Definition 3.3,

Definition 3.4

A design δ is information invariant *with respect to G and the linear regression function a, if $Q_g\mathbf{I}(\delta)Q_g' = \mathbf{I}(\delta)$ for every $g \in G$.*

Obviously, in a linear model which is linearly transformed by G the invariance of δ implies that δ is information invariant, irrespectively of the particular parametrization. We are, now, going to construct invariant designs,

Lemma 3.10

For every design δ the symmetrization $\overline{\delta} = \frac{1}{\#G}\sum_{g\in G}\delta^g$ is a design (the symmetrized design *of δ with respect to G). Moreover, every symmetrized design $\overline{\delta}$ is invariant.*

Proof

$\overline{\delta}$ is a design because the set Δ of designs on T is convex. Moreover, $\overline{\delta}(g^{-1}(B)) = \frac{1}{\#G}\sum_{\tilde{g}\in G}\delta^{\tilde{g}}(g^{-1}(B)) = \frac{1}{\#G}\sum_{\tilde{g}\in G}\delta((g\circ\tilde{g})^{-1}(B)) = \frac{1}{\#G}\sum_{\tilde{g}\in G}\delta(\tilde{g}^{-1}(B)) = \overline{\delta}(B)$ for every (measurable) subset B of T and every $g \in G$. □

Hence, a design δ coincides with its symmetrization $\overline{\delta}$ if δ is invariant. Because of $\mathbf{I}(\overline{\delta}) = \frac{1}{\#G}\sum_{g\in G}\mathbf{I}(\delta^g)$ we can also conclude that for an information invariant design δ the information matrix coincides with that of the corresponding symmetrized design, $\mathbf{I}(\delta) = \mathbf{I}(\overline{\delta})$. The concept of invariance is helpful if the transformations do not affect the value of the criterion function under consideration,

Definition 3.5

A criterion function $\Phi : \Delta \to \mathbb{R}$ *is* invariant *with respect to* G *if* $\Phi(\delta^g) = \Phi(\delta)$ *for every* $\delta \in \Delta$ *and every* $g \in G$.

Every criterion function Φ can be defined in terms of a function ϕ of the information matrix, $\Phi(\delta) = \phi(\mathbf{I}(\delta))$ (see the remark following Definition 2.1). Thus, $\Phi(\bar{\delta}) = \Phi(\delta)$ for every information invariant design δ. Of course, it depends on the group G which of the criteria will be invariant and which will not, or, more precisely, on the transformation matrices Q_g, $g \in G$, corresponding to the regression function a.

Definition 3.6

(i) *A group* G *is* orthogonal *for* a *if for every* $g \in G$ *the transformation matrix* Q_g *is orthogonal, i. e.* $Q_g' = Q_g^{-1}$.

(ii) *A group* G *is* unimodal *for* a *if for every* $g \in G$ *the transformation matrix* Q_g *is unimodal, i. e.* $|\det(Q_g)| = 1$.

We note that the orthogonality depends on a and is subject to the parametrization (see Examples 3.3 and 3.4) whereas the unimodality will not be affected by reparametrizations and we can omit the reference to the regression function a in this case. Of course, if G is orthogonal for a, then G is also unimodal.

Lemma 3.11

Let G *induce linear transformations of* a. *Then* $|\det(Q_g)| = 1$ *for every* $g \in G$.

Proof

As the parametrization given by a is minimum there exists an invariant design δ for which $\mathbf{I}(\delta)$ is regular. Hence $\mathbf{I}(\delta) = \mathbf{I}(\delta^g) = Q_g\mathbf{I}(\delta)Q_g'$ for every $g \in G$ and consequently $\det(\mathbf{I}(\delta)) = \det(\mathbf{I}(\delta))\det(Q_g)^2$ implies $|\det(Q_g)| = 1$. □

This means that, in the present setting, every group G is unimodal in contrast to invariance considerations used in the literature where the group structure is imposed on the matrices Q_g without assuming an underlying transformation group G on T.

Theorem 3.12

Let G *induce linear transformations of* a.

(i) *The* D*-criterion is invariant.*

(ii) *If* G *is orthogonal for* a, *then every* Φ_q*-criterion,* $0 \leq q \leq \infty$, *including the* A*- and* E*-criterion, is invariant.*

Proof

(i) As $|\det(Q_g)| = 1$ we have $\det(\mathbf{I}(\delta^g)) = \det(Q_g\mathbf{I}(\delta)Q_g') = \det(Q_g)^2\det(\mathbf{I}(\delta)) = \det(\mathbf{I}(\delta))$ and the D-criterion is invariant.

(ii) The eigenvalues of the information matrix are not affected by orthogonal transfromations, i. e. the eigenvalues of $\mathbf{I}(\delta)$ and $Q_g\mathbf{I}(\delta)Q_g$ coincide for orthogonal Q_g (cf RAO (1973)). As the Φ_q-criteria depend on δ only through the eigenvalues of $\mathbf{I}(\delta)$ they are invariant. □

The following straightforward result on convex criterion functions will be extremely helpful because all of our criterion functions Φ are convex, at least, after some monotonic transformations (cf the remark following Theorem 2.3),

Theorem 3.13
If $\Phi : \Delta \to I\!R$ is convex and invariant with respect to G, then $\Phi(\overline{\delta}) \le \Phi(\delta)$ for every $\delta \in \Delta$.

By Theorem 3.13 the invariant designs constitute an essentially complete class for convex and invariant criterion functions Φ. In particular, if there is an invariant design δ^* which is Φ-optimum within the class of all invariant designs, then it is also Φ-optimum within the set Δ of all designs on the design region T.

Example 3.2 *Polynomial regression (cf Example 1.2):*
Consider the model of a one-dimensional polynomial regression

$$\mu(t) = \beta_1 + \beta_2 t + ... + \beta_p t^{p-1} = \sum\nolimits_{i=1}^{p} t^{i-1}\beta_i$$

on a symmetrie design region $T = [-t_0, t_0]$. Then the sign change transformation g with $g(t) = -t$ constitutes a group $G = \{g, g_0\}$ together with the identity $g_0 = \mathrm{id}$. G induces linear transformations of the regression function a defined by $a(t) = (t^{i-1})_{i=1,...,p}$ and the corresponding transformation matrices are given by $Q_g = \mathrm{diag}(((-1)^{i-1})_{i=1,...,p})$ and $Q_{g_0} = \mathbf{E}_p$, respectively. Hence, G is orthogonal for a and the invariant designs are symmetric (for further details see Example 3.9). □

In the above Example 3.2 there exist many different invariant designs and the determination of an optimum invariant design needs further investigations with respect to the particular invariant optimality criterion under consideration. This is caused by the fact, that the group G is *small* compared to the design region T, i. e. $\{g(t);\ g \in G\}$ covers only a small portion of T for every $t \in T$.

Definition 3.7
(i) *For every $t \in T$ the subset $G(t) = \{g(t);\ g \in G\}$ of T is an* orbit *induced by the group G.*
(ii) *The group G* acts transitively *on T if $G(t) = T$ for every $t \in T$.*

It is obvious that the group G must have, at least, as many elements as the design region T if G acts transitively on T. On the contrary, if the design region T is finite, then G is a subgroup of the group of permutations and, hence, G is also finite. In this case we can characterize the invariant designs,

Lemma 3.14
Let T be finite. Then every invariant design δ is uniform on the orbits, i. e. $\delta(\{t\}) = \delta(\{g(t)\})$ for every $t \in T$ and every $g \in G$.
In particular, if G acts transitively on T, then the uniform design $\overline{\delta}$ defined by $\overline{\delta}(\{t\}) = \frac{1}{\#T}$, for every $t \in T$, is the unique invariant design with respect to G.

Proof
Let $N(t) = \#G(t)$ be the number of design points in the orbit $G(t)$. For every $t \in T$ there are $g^{(1)}, ..., g^{(N(t))} \in G$ such that $\{g^{(1)^{-1}}(t), ..., g^{(N(t))^{-1}}(t)\} = G(t)$. If δ is invariant with respect to G then $\delta(\{t\}) = \frac{1}{N(t)} \sum_{n=1}^{N(t)} \delta(\{g^{(n)^{-1}}(t)\}) = \frac{1}{N(t)}\delta(G(t))$ for every $t \in T$. In particular, this implies $\overline{\delta}(\{t\}) = \frac{1}{\#T}$ if G acts transitively on T.

Conversely, if δ is uniform on the orbits, then the invariance of δ is immediate. □

As a direct consequence of Theorem 3.13 and Lemma 3.14 we notice the following optimality of the uniform design,

Theorem 3.15

Let G induce linear transformations of a and let T be finite. If G acts transitively on T, then the uniform design $\overline{\delta}$ on T is optimum with respect to every convex, invariant criterion.

In connection with Theorem 3.12 this result establishes the Φ_q-optimality of the uniform design,

Corollary 3.16

Let G induce linear transformations of a, let T be finite, let G act transitively on T.

(i) *The uniform design $\overline{\delta}$ is D-optimum.*

(ii) *If G is orthogonal for a, then the uniform design $\overline{\delta}$ is Φ_q-optimum, $0 \leq q \leq \infty$, including the A- and E-optimality.*

Example 3.3 *One-way layout (cf Example 1.1):*

As this model is not affected by a relabeling of the levels the full permutation group G can be used as a transformation group on the set $T = \{1, ..., I\}$ of levels. The group G induces linear transformations of the regression function a and the corresponding permutation matrices $Q_g \in I\!R^{I \times I}$ are orthogonal. Moreover, the group G acts transitively on T and, hence, the uniform design $\overline{\delta}$ which assigns equal weights $\frac{1}{I}$ to each level is simultaneously A-, D- and E-optimum. Additionally, the uniform design $\overline{\delta}$ is the unique optimum design with respect to each of the A-, D- and E-criterion. □

The D-optimality will not be affected by reparametrizations, i. e. by a simultaneous, non-singular linear transformation of the regression function and the parameter vector. However, for other criteria which are not invariant with respect to every reparametrization there are different optimum designs for different parametrizations, and alternative transformation groups have to be used,

Example 3.4 *One-way layout (cf Example 1.1):*

If we are interested in the effects of the levels of one factor compared to a control level, instead of the direct effects, and in the effect of this control level, then a suitable parametrization of the one-way layout is given by $\mu(i) = \mu_0 + \alpha_i$ $i = 1, ..., I-1$, and $\mu(I) = \mu_0$. The level I is the *control level*, and the other levels are usually called the *treatment levels*. The present model is a *one-way layout with control* and the corresponding parametrization is called the *control parametrization* of the one-way layout. The regression function $a : T = \{1, ..., I\} \to I\!R^I$ is now defined by $a(i) = (1, \mathbf{e}'_{I-1,i})'$, where, again, $\mathbf{e}_{I-1,i}$ is a vector of length $I-1$ with the ith entry equal to one and all other entries equal to zero; thus, in particular, $\mathbf{e}_{I-1,I} = \mathbf{0}$. The vector $\beta = (\mu_0, \alpha_1, ..., \alpha_{I-1})$ of parameters has the following interpretation:

$\mu_0 = \mu(I)$ is the effect of the control level I;

$\alpha_i = \mu(i) - \mu(I)$ is the effect of the treatment level i compared to control.

The full group of permutations still induces linear transformations of the regression function a, but the resulting transformation matrices are not orthogonal, in general. We, thus, consider a smaller group of transformations which is suggested by the parametrization, the group G of permutations of the treatment levels. For this group the transformation matrices are given by

$$Q_g = \begin{pmatrix} 1 & \mathbf{0} \\ \mathbf{0} & \widetilde{Q}_g \end{pmatrix},$$

where $\widetilde{Q}_g$ is an $(I-1) \times (I-1)$ permutation matrix, and, hence, all Q_g are orthogonal. If we are interested in A-, D-, or E-optimality, we can confine to invariant designs $\overline{\delta}$ according to Theorem 3.13 which are uniform on the orbits (cf Lemma 3.14). In the present situation there are only two different orbits, the orbit $G(i) = \{1, ..., I-1\}$ of the treatment levels $i = 1, ..., I-1$ and the degenerate orbit $G(I) = \{I\}$ of the control level. Hence, every invariant design $\overline{\delta}$ only depends on the weight w of a single treatment level, $\overline{\delta}(\{i\}) = w$, for $i = 1, ..., I-1$, and $\overline{\delta}(\{I\}) = 1 - (I-1)w$, $0 \le w \le \frac{1}{I-1}$. In view of the identifiability of the parameters strict inequalities $0 < w < \frac{1}{I-1}$ are requested. The optimization problem is, thus, reduced to a one-dimensional real valued problem.

The invariance produces a simple form for the information matrix

$$\mathbf{I}(\overline{\delta}) = \begin{pmatrix} 1 & w\mathbf{1}'_{I-1} \\ w\mathbf{1}_{I-1} & w\mathbf{E}_{I-1} \end{pmatrix}$$

of an invariant design $\overline{\delta}$, where $\mathbf{1}_{I-1}$ denotes a vector of length $I-1$ with all its entries equal to one. By Lemma A.2 we obtain $\det(\mathbf{I}(\overline{\delta})) = w^{I-1}(1 - w(I-1))$. The determinant $\det(\mathbf{I}(\overline{\delta}))$ attains its maximum for $w^* = \frac{1}{I}$ and we recover the D-optimum design already obtained in Example 3.3.

The smallest eigenvalue of $\mathbf{I}(\overline{\delta})$ is associated with an eigenvector $(c(w), \mathbf{1}'_{I-1})'$ with its first entry $c(w) = \frac{1}{2w}(1 - w - \sqrt{(4I-3)w^2 - 2w + 1})$. The smallest eigenvalue attains its maximum $\lambda_{\min}(\mathbf{I}(\overline{\delta}^*)) = \frac{1}{4I-3}$ for the optimum weight $w^* = \frac{2}{4I-3}$ at the treatment levels which yields an E-optimum design. We add that $c(w^*) = -\frac{1}{2}$ and that for the control level the optimum weight $\delta^*(\{I\}) = \frac{2I-1}{4I-3}$ approaches $\frac{1}{2}$ if the number of treatments gets large.

For the A-criterion we compute the covariance matrix

$$\mathbf{C}(\overline{\delta}) = \begin{pmatrix} \frac{1}{1-(I-1)w} & -\frac{1}{1-(I-1)w}\mathbf{1}'_{I-1} \\ -\frac{1}{1-(I-1)w}\mathbf{1}_{I-1} & \frac{1}{w}\mathbf{E}_{I-1} + \frac{1}{1-(I-1)w}\mathbf{1}^{I-1}_{I-1} \end{pmatrix}$$

for the invarinat design $\overline{\delta}$ by means of Lemma 3.2 and the well-known inversion formula $(c_1\mathbf{E}_q - c_2\mathbf{1}^q_q)^{-1} = \frac{1}{c_1}\mathbf{E}_q + \frac{c_2}{c_1(c_1 - c_2 q)}\mathbf{1}^q_q$ for matrices of the type $c_1\mathbf{E}_q - c_2\mathbf{1}^q_q$, for $c_1 \neq 0$, $c_1 \neq c_2 q$, where $\mathbf{1}^q_q = \mathbf{1}_q\mathbf{1}'_q$ is a $q \times q$ matrix with all entries equal to one.

The trace $\mathrm{tr}(\mathbf{C}(\overline{\delta})) = \frac{I-1}{w} + \frac{I}{1-(I-1)w}$ is minimized by the weight $w^* = \frac{1}{I-1+\sqrt{I}}$ at the treatment levels which gives the A-optimum design.

In a one-way layout the weights $\delta(\{i\})$ can be recalculated from the information matrix $\mathbf{I}(\delta)$ for every level i. Hence, the A-optimum design is unique by the strict convexity of the criterion considered as a function on the information matrix. □

If the model is symmetric in its factors the permutation of factors is another important transformation group. In particular, for a two-factor model $\mu(t_1, t_2) = a(t_1, t_2)'\beta$, $t_1 \in T_1$, $t_2 \in T_2$, with $T_1 = T_2$, the permutation g is given by $g(t_1, t_2) = (t_2, t_1)$. Together with the identity $g_0 = \mathrm{id}|_{T_1\times T_2}$ the transformation g constitutes a group $G = \{g, g_0\}$. If, moreover, the regression function a is symmetric in (t_1, t_2), i. e. if for any component a_i there is a component a_j such that $a_i(t_1, t_2) = a_j(t_2, t_1)$, then G induces linear transformations of the regression function a and the transformation matrix Q_g is a permutation matrix and, hence, orthogonal.

Example 3.5 *Two-dimensional linear regression:*

For each factor the response depends linearly on the setting t_1 and t_2 respectively. If there is no interaction between the factors then the response function is given by

$$\text{(i)} \qquad \mu(t_1, t_2) = \beta_0 + \beta_1 t_1 + \beta_2 t_2$$

with regression function $a(t_1, t_2) = (1, t_1, t_2)'$. If the factors interact the response function becomes

$$\text{(ii)} \qquad \mu(t_1, t_2) = \beta_0 + \beta_1 t_1 + \beta_2 t_2 + \beta_{12} t_1 t_2$$

with an augmented regression function $a(t_1, t_2) = (1, t_1, t_2, t_1 t_2)'$. In both cases the permutation of the factors induces a permutation of the second and third component of the regression function a. □

For further applications see Sections 4 and 5 where particular families of designs are considered which are invariant with respect to the permutation of factors.

For linear aspects ψ of the unknown parameter vector use can be made of invariance considerations if the transformations are compatible with that linear aspect,

Definition 3.8

The group G induces linear transformations *of an s-dimensional linear aspect ψ if for every $g \in G$ there exists an $s \times s$ matrix $Q_{\psi,g}$ satisfying $\psi(Q_g'\beta) = Q_{\psi,g}'\psi(\beta)$ for every $\beta \in \mathbb{R}^p$.*

An s-dimensional linear aspect ψ is invariant *with respect to G if $\psi(Q_g'\beta) = \psi(\beta)$ for every $g \in G$.*

Recall that a linear aspect ψ is characterized by a selection matrix $L_\psi \in \mathbb{R}^{s\times p}$ such that $\psi(\beta) = L_\psi\beta$. Hence, G induces linear transformations of ψ if for every $g \in G$ there exists a matrix $Q_{\psi,g} \in \mathbb{R}^{s\times s}$ with $L_\psi Q_g' = Q_{\psi,g}' L_\psi$. If G induces linear transformations of an aspects then the identifiability persists under the transformations of G as follows,

Lemma 3.17

If G induces linear transformations of ψ, then $\delta \in \Delta(\psi)$ implies $\delta^g \in \Delta(\psi)$ for every $g \in G$ and, consequently, $\overline{\delta} \in \Delta(\psi)$.

Proof

For $\delta \in \Delta(\psi)$ there exists a matrix M with $L_\psi = M\mathbf{I}(\delta)$, hence $L_\psi Q_g' = M\mathbf{I}(\delta)Q_g' = MQ_g^{-1}\mathbf{I}(\delta^g)$. Now $Q_{\psi,g^{-1}}' L_\psi Q_g' = L_\psi Q_{g^{-1}}' Q_g' = L_\psi$ by the induced linear transformations and, hence, $L_\psi = Q_{\psi,g^{-1}}' M Q_{g^{-1}} \mathbf{I}(\delta^g)$ which proves the identifiability of ψ under δ^g. The identifiability of ψ under the symmetrization $\overline{\delta}$ follows from the convexity of $\Delta(\psi)$. □

It is obvious that G induces linear transformations of the whole parameter vector β by letting $Q_{\beta,g} = Q_g$ and the symmetrization preserves the identifiability of β. If

G induces linear transformations of ψ and $\delta \in \Delta(\psi)$, then the covariance matrix of δ^g can be related to the covariance matrix of δ by $\mathbf{C}_\psi(\delta^g) = L_\psi Q_g'^{-1}\mathbf{I}(\delta)^- Q_g^{-1} L_\psi' = Q'_{\psi,g^{-1}} L_\psi \mathbf{I}(\delta)^- L_\psi' Q_{\psi,g^{-1}} = Q'_{\psi,g^{-1}}\mathbf{C}_\psi(\delta) Q_{\psi,g^{-1}}$ which justifies the following definition,

Definition 3.9

Let G induce linear transformations of ψ and let $\delta \in \Delta(\psi)$. A design δ is invariant for *ψ with respect to G if $\mathbf{C}_\psi(\delta^g) = \mathbf{C}_\psi(\delta)$ for every $g \in G$.*

Hence, δ is invariant for the linear aspect ψ if $Q'_{\psi,g}\mathbf{C}_\psi(\delta)Q_{\psi,g} = \mathbf{C}_\psi(\delta)$ for every $g \in G$. In particular, δ is invariant for the whole parameter vector β if and only if δ is information invariant. If the linear aspect ψ is invariant with respect to G then every design $\delta \in \Delta(\psi)$ is invariant for ψ. Moreover, every design δ, which is invariant with respect to G, is invariant for all linear aspects ψ which can be linearly transformed by G. Additionally, if a design δ is invariant for ψ with respect to G, then its symmetrization $\overline{\delta}$ is invariant with respect to G and $\mathbf{C}_\psi(\overline{\delta}) = \mathbf{C}_\psi(\delta)$. Hence, also for linear aspects optimum designs can be found in the class of invariant designs in view of Theorem 3.13 and the following result,

Theorem 3.18

Let G induce linear transformations of ψ.

(i) *The D_ψ-criterion is invariant.*

(ii) *If $Q_{\psi,g}$ is orthogonal for every $g \in G$, then all Φ_q-criteria for ψ, $0 \le q \le \infty$, including the A_ψ- and E_ψ-criterion, are invariant.*

The proof of this theorem parallels that of the related Theorem 3.12 for the whole parameter vector β and is omitted.

Example 3.6 *One-way layout (cf Example 1.1):*

If we are interested in the treatment effects compared to control, then the control parametrization of Example 3.4 is most suitable, $\mu(i) = \mu_0 + \alpha_i$, $i = 1, ..., I-1$, and $\mu(I) = \mu_0$. The group of permutations of the treatment levels induces linear transformations $\psi(Q_g'\beta) = \widetilde{Q}_g'\psi(\beta)$ of the linear aspect ψ of the treatment effects compared to control, $\psi(\beta) = (\alpha_i)_{i=1,...,I-1}$, where $\widetilde{Q}_g$ is the corresponding $(I-1)\times(I-1)$ permutation matrix (see Example 3.4). We can confine to the invariant designs $\overline{\delta}$ with $\overline{\delta}(\{i\}) = w$, $i = 1, ..., I-1$, and $\overline{\delta}(\{I\}) = 1-(I-1)w$, where w is a weight satisfying $0 < w < \frac{1}{I-1}$. Then the uniform design $\delta^* = \frac{1}{I}\sum_{i=1}^{I}\varepsilon_i$ with equal weights $w^* = \frac{1}{I}$ is D_ψ-optimum by Lemma 3.3 and Exmaple 3.4. Because of $\mathrm{tr}(\mathbf{C}_\psi(\overline{\delta})) = (I-1)(\frac{1}{w} + \frac{1}{1-(I-1)w})$ (cf Example 3.4) the A-optimum design is obtained for the optimum weight $w^* = \frac{1}{I-1+\sqrt{I-1}}$. The maximum eigenvalue $\lambda_{\max}(\mathbf{C}_\psi(\overline{\delta})) = \frac{1}{w(1-(I-1)w)}$ of the covariance matrix of ψ is associated with the eigenvector $\mathbf{1}_{I-1}$ and the E-optimum design assigns half of the weight to the control level, $\delta^*(\{I\}) = \frac{1}{2}$, and distributes the second half equally among the treatment levels, $w^* = \frac{1}{2(I-1)}$. (Note again, that these optimum designs are unique.) □

The situation becomes more difficult if the components of the regression function a are linearly dependent on the design region T. However, in many applications such a parametrization may be more convenient to consider if it collects all the parameters of interest and if G induces linear transformations of the regression function a and the transformation matrices Q_g are more easily to handle than for a minimum parametrization. In this case we have to impose identifiability conditions in order to provide the parameters in the model equation $\mu(t) = a(t)'\beta$ with a well-defined interpretation.

Example 3.7 *Two-way layout without interactions (cf Example 1.3 (ii)):*
In this situation there are two factors of influence which can be adjusted to a finite number of levels each and there are only direct effects of these levels and no interactions between the factors, $\mu(i,j) = \mu_0 + \alpha_i^{(1)} + \alpha_j^{(2)}$, $i = 1, ..., I$, $j = 1, ..., J$. The regression function $a : \{1, ..., I\} \times \{1, ..., J\} \to \mathbb{R}^{I+J+1}$ is given by $a(i,j) = (1, \mathbf{e}'_{I,i}, \mathbf{e}'_{J,j})'$ and $\beta = (\mu_0, \alpha_1^{(1)}, ..., \alpha_I^{(1)}, \alpha_1^{(2)}, ..., \alpha_J^{(2)})' \in \mathbb{R}^{I+J+1}$ is the extended vector of parameters. Now $\mathbf{1} = \sum_{i=1}^{I} a_i = \sum_{j=1}^{J} a_{I+j}$, the components of a are linearly dependent and the dimension of the linear space spanned by the regression functions $a_1, ..., a_{I+J+1}$ is equal to $I + J - 1$.

The easiest way to reduce the number of parameters to that of the dimension of the space spanned by the regression functions is to omit the appropriate number of components of a without changing the spanned space. This is equivalent to imposing the linear constraint on β that those components of β equal to zero which belong to the omitted components of a. For example, we can request $\alpha_I^{(1)} = \alpha_J^{(2)} = 0$ which gives the parametrization of complete control, i. e. the level combination (I, J) is considered to be a control setting (see Example 5.1), or we can assume that $\mu_0 = \alpha_J^{(2)} = 0$ which is the standard-control parametrization, i. e. the parameters $\alpha_i^{(1)}$ are the direct effects of the levels of the first factor ("standard") whereas the $\alpha_j^{(2)}$ give the treatment effect of the second factor compared to a control level J (see Example 6.6). The main problem with this procedure is that the meaning of the parameters heavily depends on which components are omitted and that the parameters of real interest might not explicitly occur in this representation.

As an alternative approach we can impose directly a sufficient number of identifiability conditions on β which reduce the dimension of the parameter space to that of the space spanned by the regression funtions, but which do not change the set of possible response functions μ. In particular, the identifiability conditions $\sum_{i=1}^{I} \alpha_i^{(1)} = \sum_{j=1}^{J} \alpha_j^{(2)} = 0$ resp. $\mu_0 = \sum_{j=1}^{J} \alpha_j^{(2)} = 0$ give a standard parametrization with or without a general mean. In this case, for example, $\alpha_j^{(2)}$ is the effect of level j of the second factor compared to a general mean resp. to the mean effect of the first factor (for the latter parametrization cf Example 6.10). □

Definition 3.10
Let the dimension of span$(a_1, ..., a_p)$ *equal to* $r < p$. *A* linear identifiability condition $L_0\beta = \mathbf{0}$ *is given by an* $(p - r) \times p$ *matrix* L_0 *satisfying* rank$(L_0) = p - r$ *which preserves the set of response functions,* $\{a'\beta;\ \beta \in \mathbb{R}^p\} = \{a'\beta;\ \beta \in \mathbb{R}^p, L_0\beta = \mathbf{0}\}$.

Given a linear identifiability condition $L_0\beta = \mathbf{0}$ we can determine β uniquely from a minimum parametrization $\mu(t) = \widetilde{a}(t)'\widetilde{\beta}$, $\widetilde{a} : T \to \mathbb{R}^r$, $\widetilde{\beta} \in \mathbb{R}^r$, of the underlying model,

Lemma 3.19
Let $\mu(t) = a(t)'\beta = \widetilde{a}(t)'\widetilde{\beta}$, $t \in T$, *where* $\widetilde{a} : T \to \mathbb{R}^r$ *gives a minimum parametrization and* $a : T \to \mathbb{R}^p$, $p \geq r$.

(i) *There is a unique* $p \times r$ *matrix* L *such that* $a = L\widetilde{a}$.

(ii) *If $p > r$ and a linear identifiability condition is given by L_0, then*

$$\beta = \begin{pmatrix} L' \\ L_0 \end{pmatrix}^{-1} \begin{pmatrix} \mathbf{E}_r \\ \mathbf{0} \end{pmatrix} \widetilde{\beta} .$$

Proof

(i) This follows from the reparametrization, $\operatorname{span}(a_1, ..., a_p) = \operatorname{span}(\widetilde{a}_1, ..., \widetilde{a}_r)$, and the minimality of $\widetilde{a}$.

(ii) Because of $\widetilde{a}'\widetilde{\beta} = a'\beta = \widetilde{a}'L'\beta$ and the minimality of $\widetilde{a}$ we obtain the relation $\widetilde{\beta} = L'\beta$ and, hence,

$$\begin{pmatrix} \mathbf{E}_r \\ \mathbf{0} \end{pmatrix} \widetilde{\beta} = \begin{pmatrix} L' \\ L_0 \end{pmatrix} \beta .$$

Now $\operatorname{rank}(L) = r$, $\operatorname{rank}(L_0) = p - r$, and the rows of L' and L_0 are independent. Thus, the $p \times p$ matrix $(L|L_0')$ is regular which proves the representation. □

We note that Lemma 3.19 yields the identifiability of β in the model under the identifiability condition $L_0\beta = \mathbf{0}$ and presents, additionally, a constructive way how to calculate β from a solution $\widetilde{\beta}$ under a minimum parametrization. In particular, β is a linear aspect of $\widetilde{\beta}$.

Definition 3.11

The group G induces linear transformations *of a linear identifiability condition $L_0\beta = \mathbf{0}$ if for every $g \in G$ there exists an $(p-r) \times (p-r)$ matrix $Q_{0,g}$ satisfying $Q_{0,g}' L_0 = L_0 Q_g'$.*

A linear identifiability condition $L_0\beta = \mathbf{0}$ is invariant *with respect to G, if $L_0 Q_g' = L_0$ for every $g \in G$.*

The next result links the identifiability condition with the invariance considerations,

Lemma 3.20

Let G induce linear transformations of a and of the linear identifiability condition $L_0\beta = \mathbf{0}$. Then $\mathbf{C}_\beta(\delta^g) = Q_{g^{-1}}' \mathbf{C}_\beta(\delta) Q_{g^{-1}}$ for every $g \in G$ and and for every $\delta \in \Delta(\beta)$.

Proof

Let a minimum parametrization $\mu(t) = \widetilde{a}(t)'\widetilde{\beta}$ be given by $\widetilde{a} : T \to I\!R^r$. Then $\beta = L_\beta \widetilde{\beta}$ can be written as a linear aspect of $\widetilde{\beta}$ where $a = L\widetilde{a}$ and L_β is defined in Lemma 3.19. Hence, $\widetilde{a} = L_\beta' a$ and G induces linear transformations of $\widetilde{a}$ with transformation matrices $\widetilde{Q}_g = L_\beta' Q_g L$. Then $C_\beta(\delta) = L_\beta C_{\widetilde{\beta}}(\delta) L_\beta'$ and $C_\beta(\delta^g) = L_\beta C_{\widetilde{\beta}}(\delta^g) L_\beta' = L_\beta \widetilde{Q}_{g^{-1}}' C_{\widetilde{\beta}}(\delta) \widetilde{Q}_{g^{-1}} L_\beta'$ for every $g \in G$.

It remains to show that $L_\beta' Q_g = \widetilde{Q}_g L_\beta'$ for every $g \in G$ and, hence, for g^{-1}. Now $L_\beta' Q_g - \widetilde{Q}_g L_\beta' = L_\beta' Q_g ((L|L_0')(L|L_0')^{-1} - L L_\beta') = L_\beta' Q_g (\mathbf{0}|L_0')(L|L_0')^{-1}$ and, because of the invariance of L_0, we see that $L_0 Q_g' L_\beta = Q_{0,g}' L_0 L_\beta = \mathbf{0}$, where the last equality is due to the identifiability condition $L_0\beta = L_0 L_\beta \widetilde{\beta} = \mathbf{0}$ for every $\widetilde{\beta}$ which, finally, proves the assertion. □

If G is orthogonal then the present Lemma 3.20 together with Theorem 3.13 ensures that the invariant designs constitute an essentially complete class for every Φ_q-criterion, $0 < q \leq \infty$, based on $C_\beta(\delta)$,

Theorem 3.21

Let G induce linear transformations of a and of the linear identifiability condition $L_0\beta = \mathbf{0}$ and let G be orthogonal for a. Then $\Phi_q(\overline{\delta}) \leq \Phi_q(\delta)$ for all $q > 0$ and every $\delta \in \Delta$, where $\overline{\delta}$ is the symmetrization of δ.

In particular, the uniform design $\overline{\delta}$ is Φ_q-optimum, $0 < q \leq \infty$, if G acts transitively on T. The result of Theorem 3.21 cannot be extended directly to the D-criterion ($q = 0$) if the model is overparametrized because then $\det(\mathbf{I}(\delta)) = 0$, the covariance matrix $C_\beta(\delta)$ is singular for every $\delta \in \Delta(\beta)$ and the D-criterion is no longer meaningful. However, D-optimality of invariant designs can be obtained for every minimum parametrization,

Theorem 3.22

If there exists a linear parametrization $\mu(t) = \bar{a}(t)'\bar{\beta}$ and a linear identifiability condition $L_0\beta = \mathbf{0}$, such that G induces linear transformations of $\bar{a}$ and of $L_0\bar{\beta} = \mathbf{0}$ and let G be orthogonal for a, then $\det(\int aa'd\overline{\delta}) \geq \det(\int aa'd\delta)$ for every minimum linear parametrization $\mu(t) = a(t)'\beta$ and for every $\delta \in \Delta$, where $\overline{\delta}$ is the symmetrization of δ.

In particular, in case G acts transitively on T the uniform design $\overline{\delta}$ is D-optimum if there exists a linear parametrization $\mu(t) = \bar{a}(t)'\bar{\beta}$ with corresponding identifiability condition $L_0\beta = \mathbf{0}$ satisfying the assumptions of Theorem 3.22.

Example 3.8 *One-way layout (cf Example 1.1):*

Similar to Example 3.7 we consider the one-way layout in standard parametrization with a general mean $\mu(i) = \mu_0 + \alpha_i$, $i = 1, ..., I$, with identifiability condition $\sum_{i=1}^{I} \alpha_i = 0$. Then $a(i) = (1, \mathbf{e}'_{I,i})'$ and the transformation matrices with respect to the full group G of permutations are given by

$$Q_g = \begin{pmatrix} 1 & \mathbf{0} \\ \mathbf{0} & \widetilde{Q}_g \end{pmatrix},$$

where $\widetilde{Q}_g$ is the corresponding $I \times I$ permutation matrix. Hence, Q_g is orthogonal and the identifiability condition $L_0 = (0, \mathbf{e}'_{I,i})$ is invariant. As G acts transitively on T the uniform design $\overline{\delta}$ is A- and E-optimum for the extended parameter vector $\beta = (\mu_0, \alpha_1, ..., \alpha_I)'$. □

Further applications will be given in Examples 6.9 and 6.10.

Example 3.9 *Polynomial regression (cf Example 1.2):*

We want to determine optimum designs in linear resp. quadratic regression on the standardized design region $T = [-1, 1]$. By $m_j(\delta) = \int t^j\,\delta(dt)$ we denote the jth *moments* of the design δ which are the entries in the corresponding information matrix $\mathbf{I}(\delta)$. As we consider optimality criteria which are invariant with respect to sign change (cf Example 3.2) we can confine to symmetric designs $\overline{\delta}$ for which the odd moments vanish, $m_j(\overline{\delta}) = 0$ for odd j.

For linear regression

$$\mathbf{I}(\overline{\delta}) = \begin{pmatrix} 1 & 0 \\ 0 & m_2(\overline{\delta}) \end{pmatrix}$$

is the information matrix of a symmetric design $\overline{\delta}$. Now, $m_2(\overline{\delta}) \leq 1$ and equality is attained for the design δ^* which assigns equal weights $\frac{1}{2}$ to each of the endpoints -1 and

1 of the interval T. Hence, $\mathbf{I}(\overline{\delta}) \leq \mathbf{I}(\delta^*)$ and δ^* is optimum for every criterion which is invariant with respect to sign change, including A-, D- and E-optimality (cf Example 3.1).

For quadratic regression

$$\mathbf{I}(\overline{\delta}) = \begin{pmatrix} 1 & 0 & m_2(\overline{\delta}) \\ 0 & m_2(\overline{\delta}) & 0 \\ m_2(\overline{\delta}) & 0 & m_4(\overline{\delta}) \end{pmatrix}$$

is the information matrix of a symmetric design $\overline{\delta}$. Now, $m_4(\overline{\delta}) \leq m_2(\overline{\delta})$ because of $t^4 \leq t^2$ on $T = [-1, 1]$ and equality is attained for those designs $\delta^{(w)}$ which assign equal weights w to each of the endpoints -1 and 1 of the interval T and the remaining weight $1 - 2w$ to the midpoint 0. Hence, $\mathbf{I}(\overline{\delta}) \leq \mathbf{I}(\delta^{(w)})$ where $w = \frac{1}{2}m_2(\overline{\delta})$, and for every criterion which is invariant with respect to sign change an optimum design can be found in the class of symmetric designs $\delta^{(w)}$ supported on $\{-1, 0, 1\}$, $0 < w < \frac{1}{2}$. It remains to determine the optimum weights w^* for the different criteria. The determinant $\det(\mathbf{I}(\delta^{(w)})) = (2w)^2(1 - 2w)$ attains its maximum at $w^* = \frac{1}{3}$ and the D-optimum design assigns equal weights $\frac{1}{3}$ to each of the supporting points. (Note that $\delta^* = \delta^{(w^*)}$ is concentrated on a minimum support and, hence, the weights have to be equal by Lemma 2.11.) The minimum eigenvalue $\lambda_{\min}(\mathbf{I}(\delta^{(w)})) = \frac{1}{2}(1 + 2w - \sqrt{1 - 4w + 20w^2})$ attains its maximum at $w^* = \frac{1}{5}$ and the E-optimum design assigns equal weights $\frac{1}{5}$ to each of the endpoints of the interval and the remaining weight $\frac{3}{5}$ to the midpoint. As

$$\mathbf{C}(\delta^{(w)}) = \frac{1}{2w(1-2w)} \begin{pmatrix} 2w & 0 & -2w \\ 0 & 1 - 2w & 0 \\ -2w & 0 & 1 \end{pmatrix}$$

the trace $\operatorname{tr}(\mathbf{C}(\delta^{(w)})) = \frac{1}{w(1-2w)}$ of the covariance matrix attains its minimum at $w^* = \frac{1}{4}$ and the A-optimum design assigns equal weights $\frac{1}{4}$ to each of the endpoints of the interval and the remaining weight $\frac{1}{2}$ to the midpoint.

Note that the results on D-optimality are valid for general intervals $T = [\tau_1, \tau_2]$ because translations and scale transformations result in linear transformations of the regression functions in polynomial regression and, hence, do not affect the D-optimality.

The present majorization technique can be generalized to find optimum designs for polynomial regression with a minimum support. For more details concerning the moments $m_j(\delta)$, their relationship to the ordering of information matrices, and the impact on the specification of optimum designs we refer to KARLIN and STUDDEN (1966) and GIOVAGNOLI, PUKELSHEIM, and WYNN (1987). □

We add a technical result which will be helpful in later applications,

Lemma 3.23

Let $\gamma = \int a'\, d\delta (\int aa'\, d\delta)^- \int a\, d\delta$.

(i) *If* $\mathbf{1} \in \operatorname{span}(a_1, ..., a_p)$, *then* $\gamma = 1$.

(ii) *Let G induce linear transformations of a, let G act transitively on T and let $\overline{\delta}$ be the uniform design on T. If* $\mathbf{1} \notin \operatorname{span}(a_1, ..., a_p)$, *then* $\int a\, d\overline{\delta} = \mathbf{0}$ *and, consequently,* $\gamma = 0$.

Proof

(i) As $\mathbf{1} \in \text{span}(a_1, ..., a_p)$ there exists a vector $\ell \in \mathbb{R}^p$ such that $\ell' a(t) = 1$ for every $t \in T$ and, hence, $\gamma = \ell' \int aa'\, d\delta \,(\int aa'\, d\delta)^- \int aa'\, d\delta\, \ell = \ell' \int aa'\, d\delta\, \ell = 1$.

(ii) First, we assume a minimum parametrization. If $\mathbf{1} \notin \text{span}(a_1, ..., a_p)$ the matrix

$$J = \begin{pmatrix} 1 & \int a'\, d\bar{\delta} \\ \int a\, d\bar{\delta} & \int aa'\, d\bar{\delta} \end{pmatrix}$$

is regular because J is the information matrix of the uniform design $\bar{\delta}$ in the *extended* model $\mu^{(0)}(t) = \beta_0 + a(t)'\beta$, $t \in T$. Now, $\det(J) = (1-\gamma)\det(\int aa'\, d\bar{\delta})$ by the formula for the determinant of partitioned matrices (Lemma A.2) which, together with $\det(J) > 0$, yields $\gamma \neq 1$.

Because of the invariance of the uniform design $\bar{\delta}$ we obtain $\int a'\, d\bar{\delta}\,(\int aa'\, d\bar{\delta})^{-1} a(t) = \int a'\, d\bar{\delta}\, Q_g' Q_g'^{-1} (\int aa'\, d\bar{\delta})^{-1} Q_g^{-1} Q_g a(t) = \int a'\, d\bar{\delta}\,(\int aa'\, d\bar{\delta})^{-1} a(g(t))$ for every transformation g and, hence, the function $\int a'\, d\bar{\delta}\,(\int aa'\, d\bar{\delta})^{-1} a(t) = \gamma$ is constant in t. Thus, $\gamma \int a'\, d\bar{\delta} = \int a'\, d\bar{\delta} (\int aa'\, d\bar{\delta})^{-1} \int aa'\, d\bar{\delta} = \int a'\, d\bar{\delta}$ in view of Lemma 3.1 which proves $\int a\, d\bar{\delta} = \mathbf{0}$ as $\gamma \neq 1$.

For a parametrization which is not minimum there is an associated minimum parametrization given by a regression function $\tilde{a}$ with $a = L\tilde{a}$ (see Lemma 3.19). The matrix L can be augmented by linearly independent columns to a regular matrix $\bar{L}$ such that $a = \bar{L}(\tilde{a}', \mathbf{0})'$, and the result follows from the minimum parametrization case. □

As already mentioned the assumption of G being finite may be replaced by the requirement that G is a compact group. If G acts transitively on T, then the uniform design $\bar{\delta}$ on T is the unique invariant design (for details see HALMOS (1974) or EATON (1989)). The following example illustrates this concept,

Example 3.10 *Trigonometric regression of degree M:*

$$\mu(t) = \beta_0 + \sum\nolimits_{m=1}^{M} (\beta_{m1}\sin(mt) + \beta_{m2}\cos(mt)) \, ,$$

$t \in T = [0, 2\pi)$. The unknown parameter $\beta = (\beta_0, \beta_{11}, \beta_{12}, ..., \beta_{M2})'$ has dimension $p = 2M+1$ and the regression function $a : T \to \mathbb{R}^p$ is given by $a(t) = (1, \sin(t), \cos(t), ..., \sin(Mt), \cos(Mt))'$. For the group G acting on T we consider the group of all translations modulo 2π, $G = \{g_y;\ g_y(t) = t - y \bmod 2\pi, y \in [0, 2\pi)\}$, where $t - y \bmod 2\pi = t - y \in T$ if $y \le t$ and $t - y \bmod 2\pi = t - y + 2\pi \in T$ if $y > t$. G induces linear transformations of the regression function a with transformation matrices

$$Q_{g_y} = \begin{pmatrix} 1 & \mathbf{0} \\ \mathbf{0} & \text{diag}\left(\begin{pmatrix} \cos(my) & -\sin(my) \\ \sin(my) & -\cos(my) \end{pmatrix}_{m=1,...,M} \right) \end{pmatrix}$$

which are orthogonal. The group G acts transitively on T. Moreover, G is a compact group (mod 2π) and the uniform design $\bar{\delta}$ on T is the normalized LEBESGUE measure $\bar{\delta} = \frac{1}{2\pi}\lambda|_T$ restricted to the set $T = [0, 2\pi)$. The information matrix

$$\mathbf{I}(\bar{\delta}) = \begin{pmatrix} 1 & \mathbf{0} \\ \mathbf{0} & \frac{1}{2}\mathbf{E}_{2M} \end{pmatrix} .$$

of $\overline{\delta}$ is diagonal because of the orthogonality of the regression functions with respect to the uniform design $\overline{\delta}$.

We note that every uniform design on an equidistant grid with at least $p = 2M + 1$ supporting points leads to the same information matrix. More precisely, let $\delta^{(N,\tau)} = \frac{1}{N}\sum_{n=1}^{N} \epsilon_{\tau+\frac{n-1}{N}2\pi}$ be the uniform design on the set of equidistant supporting points $\{\tau + \frac{n-1}{N}2\pi;\ n = 1,...,N\}$, $0 \leq \tau < \frac{1}{N}2\pi$. If $N \geq p = 2M + 1$, then $\mathbf{I}(\delta^{(N,\tau)}) = \mathbf{I}(\overline{\delta})$ and $\delta^{(N,\tau)}$ is an optimum design with respet to each invariant criterion (cf PUKELSHEIM (1980), for D-optimality see HOEL (1965)). □

Part II

Particular Classes of Multi-factor Models

Most experimental situations are governed by a number of different factors of influence which may be related by various interaction structures. The construction of a good or even optimum design is substantially more complicated in such a situation than for models with a single factor of influence for which the theory of optimum designs is well developed. Therefore, it is desirable to reduce the complex design problem in a multi-factor model to its one-dimensional counterparts and, then, use the known results and techniques for single factor models.

The importance of optimum design in multi-factor models has become prominent by the increasing interest in the new philosophy of quality design related to the name of TAGUCHI (1987) which was advanced and clarified by BASSO, WINTERBOTTOM, and WYNN (1986) and in the monograph by LOGOTHETIS and WYNN (1989).

For *analysis of variance models* in which all factors of influence are qualitative optimum experimental design has been treated extensively in the literature (see e. g. KUROTSCHKA (1971–1978), GAFFKE and KRAFFT (1979), GAFFKE (1981) and PUKELSHEIM (1983); for a review see also SHAH and SINHA (1989)). Also some results have been obtained for multivariate polynomial regression models in which all factors are quantitative (see e. g. FARRELL, KIEFER and WALBRAN (1967) and VUCHKOV, YONTCHEV, and DAMGALIEV (1983)).

HARVILLE (1975) is the first source for optimum designs in situations where both qualitative and quantitative factors influence the outcome of the experiment. In the sequel, COX (1984), ATKINSON (1988), DRAPER and JOHN (1988) and ATKINSON and DONEV (1992) considered the treatment of models with different types of factors as one of the main topics of further research in the field of experimental design. Those more complex models with both qualitative and quantitative factors have been investigated by LOPES TROYA (1982), KUROTSCHKA (1984–1988), KUROTSCHKA and WIERICH (1984), WIERICH (1984–1989) BUONACCORSI and IYER (1986), LIM, STUDDEN, and WYNN (1988), DONEV (1989), KUROTSCHKA, SCHWABE and WIERICH (1992) and DAVID (1994), besides others.

The approach which we will choose here is independent of the actual structure of the influence of the single factors and, hence, covers models with both qualitative and quantitative factors as well as purely qualitative or purely quantitative models (cf also

RAFAJŁOWICZ and MYSZKA (1988, 1992)). This approach can successfully be extended to experimental situations with multivariate observations (cf KUROTSCHKA and SCHWABE (1995)). As has been pointed out by KUROTSCHKA (1984) every multi-factor model

$$\mu(t_1, ..., t_K) = \sum_{i=1}^{p} a_i(t_1, ..., t_K)\beta_i = a(t_1, ..., t_K)'\beta$$

can be conditioned on any of its factors (cf also WIERICH (1986c)). Because different factors can be subsumed into a new meta-factor it is sufficient to illustrate this procedure for a two-factor model

$$\mu(t_1, t_2) = \sum_{i=1}^{p} a_i(t_1, t_2)\beta_i = a(t_1, t_2)'\beta \ ,$$

$(t_1, t_2) \in T$. Define the marginal design region $T_1 = \{t_1;\ (t_1, t_2) \in T \text{ for some } t_2\}$ by the projection of T onto the first component. Then for every $t_1 \in T_1$ we obtain a *conditional model*

$$\mu_{2|t_1}(t_2) = \sum_{j=1}^{p_2(t_1)} a_j^{(2|t_1)}(t_2)\beta_j^{(2|t_1)} = a^{(2|t_1)}(t_2)'\beta^{(2|t_1)} \ ,$$

$t_2 \in T_2(t_1)$, where $T_2(t_1) = \{t_2;\ (t_1, t_2) \in T\}$ is the t_1-cut of T, $(a_j^{(2|t_1)})_{j=1,...,p_2(t_1)}$ is a set of linearly independent regression functions which span $(a_i(t_1, \cdot))_{i=1,...,p}$, and $\beta^{(2|t_1)} \in I\!R^{p_2(t_1)}$ is a parameter vector which may depend on t_1. This dependence has to be specified and leads to different classes of models. The present ideas which are rather abstract in their generality have to be adapted to the particular models considered.

As we will be mainly interested in situations for which we can construct product designs which are optimum the design region and the regression function should reflect this product structure. So in Sections 4 to 6 we assume that $T_2(t_1) = T_2$ and $a^{(2|t_1)} = a^{(2)}$ are independent of t_1. The resulting model $\mu_2(t_2) = a^{(2)}(t_2)'\beta^{(2)}$, $t_2 \in T_2$, will be called the *marginal model* for the second factor. The crucial difference between the models will be the dependence structure of $\beta^{(2|t_1)}$ on t_1.

First assume, to clarify these ideas, that the factor on which we condition is qualitative which means that the corresponding design region T_1 is finite. If the conditional parameter vector $\beta^{(2|t_1)}$ does not depend on t_1, then we can ignore the first factor because the response function μ will be constant in t_1. Disregarding this degenerate case we will be concerned with the situations that the parameter vector $\beta^{(2|t_1)}$ depends either completely on t_1 or that a prespecified part of it depends on t_1.

In the first case of $\beta^{(2|t_1)}$ depending completely on t_1 we consider the model

$$\begin{aligned} \mu(t_1, t_2) &= \sum_{j=1}^{p_2} a_j^{(2)}(t_2)\beta_{t_1,j} \\ &= \sum_{i \in T_1} \sum_{j=1}^{p_2} a_i^{(1)}(t_1)a_j^{(2)}(t_2)\beta_{i,j} \ , \end{aligned}$$

$(t_1, t_2) \in T_1 \times T_2$, with indicator functions $a^{(1)}$ ($a_i^{(1)}(t_1) = 1$ if $t_1 = i$ and $a_i^{(1)}(t_1) = 0$ if $t_1 \neq i$). This constitutes a model with complete product-type interactions as treated in Section 4. The model

$$\mu_1(t_1) = \sum_{i \in T_1} a_i^{(1)}(t_1)\beta_i^{(1)} = a^{(1)}(t_1)'\beta^{(1)} \ ,$$

$t_1 \in T_1$, will be called the *marginal model* associated with the first factor. Note that μ_1 also describes the conditional model of μ given the value t_2 for the second factor.

If $\mu_{2|t_1}(t_2) = \beta_{t_1,0}+\sum_{j=2}^{p_2} a_j^{(2)}(t_2)\beta_j$, i. e. if only the constant (or intercept) term depends on t_1, then

$$\begin{aligned}\mu(t_1,t_2) &= \beta_{t_1,0} + \sum_{j=2}^{p_2} a_j^{(2)}(t_2)\beta_j \\ &= \sum_{i\in T_1} a_i^{(1)}(t_1)\beta_{i,0} + \sum_{j=2}^{p_2} a_j^{(2)}(t_2)\beta_j \ ,\end{aligned}$$

$(t_1,t_2) \in T_1 \times T_2$, constitutes a model without interactions between the factors as considered in Section 5.

Finally, if a substantial part β_1 of β depends on t_1 and another substantial part β_0 does not, then

$$\begin{aligned}\mu(t_1,t_2) &= \sum_{j=1}^{p_{2,1}} a_j^{(2)}(t_2)\beta_{t_1,j} + \sum_{j=p_{2,1}+1}^{p_2} a_j^{(2)}(t_2)\beta_{0,j} \\ &= \sum_{i\in T_1} \sum_{j=1}^{p_{2,1}} a_i^{(1)}(t_1)a_j^{(2)}(t_2)\beta_{i,j} + \sum_{j=p_{2,1}+1}^{p_2} a_j^{(2)}(t_2)\beta_{0,j} \ ,\end{aligned}$$

$(t_1,t_2) \in T_1 \times T_2$. Such models with partly interacting factors will be treated in Subsection 6.2.

The qualitative structure of the first factor can be replaced by the requirement that the dependence of $\beta^{(2|t_1)}$ on t_1 is linear, i. e. $\beta_j^{(2|t_1)} = \sum_{i=1}^{p_1} a_i^{(1)}(t_1)\beta_{ij}^{(1)}$, for $j = 1, ..., p_2$, in case of complete product-type interactions (see Sections 4) and $\beta_1^{(2|t_1)} = \sum_{i=1}^{p_1} a_i^{(1)}(t_1)\beta_{i1}^{(1)}$ in case of no interactions (see Sections 5) where $\beta_j^{(2|t_1)}$ is constant in t_1 for $j = 2, ..., p_2$.

A generalization to complete interaction structures in the presence of more than two factors is presented in Subsection 6.1, and some results are briefly mentioned for other interaction structures and different design regions in the final Section 7.

4 Complete Product-type Interactions

In the present and the following sections we deal with different interaction structures in models where more than one factor of interest is present. At first we treat the case of complete interactions which has been thoroughly investigated in the literature starting from HOEL (1965). With respect to the methods of proof involved the present section is dedicated to the equivalence theorems which have been presented in Section 2.

We start with the introduction of the marginal single factor models described by their corresponding marginal response functions

$$\mu_k(t_k) = a^{(k)}(t_k)'\beta^{(k)} , \tag{4.1}$$

$t_k \in T_k$, $a^{(k)} : T_k \to \mathbb{R}^{p_k}$, $\beta^{(k)} \in \mathbb{R}^{p_k}$, $k = 1, ..., K$. The resulting K-factor model with complete interactions is defined by the response function

$$\mu(t_1, ..., t_K) = \sum\nolimits_{i_1=1}^{p_1} \cdots \sum\nolimits_{i_K=1}^{p_K} a_{i_1}^{(1)}(t_1) \cdot ... \cdot a_{i_K}^{(K)}(t_K)\beta_{i_1,...,i_K} , \tag{4.2}$$

$(t_1, ..., t_K) \in T = T_1 \times ... \times T_K$, in which all K-fold products of those components $a_i^{(k)}$ occur which belong to the regression functions $a^{(k)}$ associated with the different factors. This may be rewritten in a more comprehensive way by making use of the notation of KRONECKER products

$$\mu(t_1, ..., t_K) = (a^{(1)}(t_1) \otimes ... \otimes a^{(K)}(t_K))' \beta = \left(\bigotimes\nolimits_{k=1}^{K} a^{(k)}(t_k)\right)' \beta ,$$

$(t_1, ..., t_K) \in T = T_1 \times ... \times T_K$, $\beta \in \mathbb{R}^p$, with $p = \prod_{k=1}^{K} p_k$ (For the notation of KRONECKER products and the rules of how to do calculus with them we refer e. g. to GRAHAM (1981), RAO (1973), or MAGNUS and NEUDECKER (1988) besides others). Hence, β collects the unknown parameters $\beta_{i_1,...,i_K}$, $i_k = 1, ..., p_k$, $k = 1, ..., K$, in lexicographic order $\beta = (\beta_{1,...,1,1}, \beta_{1,...,1,2}, ..., \beta_{1,...,1,p_K}, \beta_{1,...,2,1}, ..., \beta_{p_1,...,p_{K-1},p_K})$ and the response function

$$\mu(t_1, ..., t_K) = a(t_1, ..., t_K)'\beta$$

is parametrized by the regression function $a : T_1 \times T_2 \times ... \times T_K \to R^p$ with $a(t_1, ..., t_K) = \bigotimes_{k=1}^{K} a^{(k)}(t_k)$.

In particular, we will be interested in the two-factor model with complete product-type interactions

$$\mu(t_1, t_2) = (a^{(1)}(t_1) \otimes a^{(2)}(t_2))'\beta , \tag{4.3}$$

$(t_1, t_2) \in T = T_1 \times T_2$, $\beta \in \mathbb{R}^{p_1 p_2}$, from which results concerning K-factor models can be obtained by induction in the very same way as e. g. K-fold products can be calculated by induction from the product of two terms.

Example 4.1 *Two-way layout with complete interactions:*

Starting from the one-way layout models $\mu_1(i) = \alpha_i^{(1)}$, $i = 1..., I$, and $\mu_2(j) = \alpha_j^{(2)}$, $j = 1..., J$, in standard parametrization (without general mean, cf Example 3.3) we obtain the two-way layout with complete interactions

$$\mu(i, j) = \alpha_{ij} ,$$

$i = 1...,I$, $j = 1,...,J$, in standard parametrization (without general mean, cf Example 1.3 (i)). The marginal regression functions are given by $a^{(1)}(i) = \mathbf{e}_{I,i}$ and $a^{(2)}(j) = \mathbf{e}_{J,j}$. We recall that e. g. $\mathbf{e}_{I,i}$ is a vector of length I with the ith entry equal to one and all other entries equal to zero. Thus $a(i,j) = \mathbf{e}_{I,i} \otimes \mathbf{e}_{J,j} = \mathbf{e}_{IJ,(i-1)J+j}$ and we notice that this model may be identified with a one-way layout with IJ levels after renaming the parameters (cf KUROTSCHKA (1984)). Hence, the equireplicated design $\overline{\delta}$ which assigns equal weights $\frac{1}{IJ}$ to each level combination is simultaneously D-, A- and E-optimum (see Example 3.3). □

Example 4.2 *Polynomial regression:*
The underlying model is described by

$$\mu(t_1, t_2) = \sum_{i=1}^{p_1} \sum_{j=1}^{p_2} t_1^{i-1} t_2^{j-1} \beta_{ij} ,$$

$(t_1, t_2) \in T = T_1 \times T_2$, with corresponding marginal models $\mu_1(t_1) = \sum_{i=1}^{p_1} t_1^{i-1} \beta_i^{(1)}$, $t_1 \in T_1$, and $\mu_2(t_2) = \sum_{j=1}^{p_2} t_2^{j-1} \beta_j^{(2)}$, $t_2 \in T_2$. In particular, for linear marginals ($p_1 = p_2 = 2$) we consider the model of two-dimensional linear regression with interactions (cf Example 3.5 (ii))

$$\text{(i)} \qquad \mu(t_1, t_2) = \beta_0 + \beta_1 t_1 + \beta_2 t_2 + \beta_{12} t_1 t_2$$

and for quadratic marginals ($p_1 = p_2 = 3$) we consider the model

$$\text{(ii)} \qquad \begin{aligned} \mu(t_1, t_2) \;&=\; \beta_0 + \beta_1 t_1 + \beta_2 t_2 + \beta_{11} t_1^2 + \beta_{22} t_2^2 + \beta_{12} t_1 t_2 \\ &\quad + \beta_{112} t_1^2 t_2 + \beta_{122} t_1 t_2^2 + \beta_{1122} t_1^2 t_2^2 , \end{aligned}$$

after a natural relabeling of the parameters. Note that the latter model includes more parameters than the commonly used model of polynomial regression up to degree 2 (see Example 7.4). □

Example 4.3 *Intra-class models with identical partial models:*
The underlying model has a qualitative part which can be described by a one-way layout $\mu_1(i) = \alpha_i$, $i = 1,...,I$, and a quantitative component $\mu_2(u) = f(u)'\beta^{(2)}$, $u \in U$. (Typically $f(u) = (1, u, u^2, ..., u^{p_2-1})'$ such that μ_2 is a polynomial in u.) Then

$$\mu(i, u) = f(u)'\beta_i ,$$

$(i, u) \in \{1, ..., I\} \times U$, is an intra-class model with identical partial model $\mu_{2|i}$ in each class i (see e. g. SEARLE (1971), pp 355, KUROTSCHKA (1984), and BUONACCORSI and IYER (1986)), $\beta_i \in \mathbb{R}^{p_2}$, $\beta = (\beta_1', ..., \beta_I')'$. □

Throughout we denote the set of all designs for the marginal models (4.1) by Δ_k, $k = 1, ..., K$, and for the whole model with complete product-type interactions (4.2) or (4.3) by Δ. Then $\Delta_k(\psi_k)$ and $\Delta(\psi)$ are those designs for which the linear aspects ψ_k and ψ are identifiable within the model (4.1), $k = 1, ..., K$, or (4.2), (4.3), respectively. In particular, $\Delta_k(\beta^{(k)})$ and $\Delta(\beta)$ are the designs which makes the whole parameter vector $\beta^{(k)}$ resp. β identifiable. The transformation matrices associated with the linear aspects

ψ_k and ψ in the appropriate models will be denoted by L_{k,ψ_k} and L_ψ, respectively. In the same spirit $\mathbf{I}_k(\delta_k) = \int a^{(k)}a^{(k)\prime}\,d\delta_k$ and $\mathbf{I}(\delta) = \int aa'\,d\delta$ are the information matrices of $\delta_k \in \Delta_k$ and $\delta \in \Delta$, and $\mathbf{C}_{k,\psi_k}(\delta_k)$ and $\mathbf{C}_\psi(\delta)$ are the covariance matrices for ψ_k under δ_k and for ψ under δ, respectively. In particular, $\mathbf{C}_k(\delta_k)$ and $\mathbf{C}(\delta)$ are the covariance matrices for the whole parameter vectors $\beta^{(k)}$ under δ_k and for β under δ. Similar notations will be used in the subsequent sections.

Example 4.4 *Two-way layout with complete interactions (cf Example 4.1):*
The information matrix $\mathbf{I}(\delta)$ is diagonal with its entries equal to the weights $\delta(\{(i,j)\})$ of each level combination listed in lexicographic order. □

Example 4.5 *Intra-class models with identical partial models (cf Example 4.3):*
In this model the information matrix $\mathbf{I}(\delta) = \mathrm{diag}((\delta_1(\{i\})\mathbf{I}_{2|i}(\delta_2(i,\cdot)))_{i=1,...,I})$ is block diagonal where the matrices $\mathbf{I}_{2|i}(\delta_2(i,\cdot)) = \int f(u)f(u)'\delta_2(i,du)$ involved in the blocks are the information matrices in the second marginal model with respect to the conditional design $\delta_2(i,\cdot)$ given that the first factor equals i, i.e. that we are observing in class i. In terms of measure theory δ_2 is a MARKOV kernel and $\delta = \delta_1 \otimes \delta_2$ is the product of the marginal design δ_1 on the first marginal model and the conditional design on the second marginal model given by the MARKOV kernel δ_2. To be more specific, for δ supported on a finite set, $\delta(\{(i,u)\}) = \delta_1(\{i\})\delta_2(i,\{u\})$ is the weight for observing at $(i,u) \in \{1,...,I\} \times U$ (cf Section 7 for these considerations in the framework of more general models). □

Our aim is to demonstrate that for most problem specifications the optimum design can be constructed as a product design in the present model of complete product-type interactions. Moreover, the marginals of those optimum designs are formed by the optimum designs for the corresponding marginal models. (Although product designs $\delta_1 \otimes \delta_2$ are defined as product measures no further knowledge of measure theory is requested because the marginals can be chosen to be supported on a finite set each (see the remark at the end of Section 1). Hence, the product design $\delta_1 \otimes \delta_2$ is supported on the cross-product of those finite sets which is again finite and the weights are given by the product of their marginals $\delta_1 \otimes \delta_2(\{(t_1,t_2)\}) = \delta_1(\{t_1\})\delta_2(\{t_2\})$.)

For a product design the information matrix and, hence, the covariance matrix factorizes into its marginal counterparts:

Lemma 4.1
(i) *For every $\delta_1 \in \Delta_1$ and $\delta_2 \in \Delta_2$*

$$\mathbf{I}(\delta_1 \otimes \delta_2) = \mathbf{I}_1(\delta_1) \otimes \mathbf{I}_2(\delta_2)\ .$$

(ii) *Let ψ be a linear aspect of β for which the associated transformation matrix factorizes, $L_\psi = L_{1,\psi_1} \otimes L_{2,\psi_2}$, and let $\delta_k \in \Delta_k(\psi_k)$. Then $\delta_1 \otimes \delta_2 \in \Delta(\psi)$ and*

$$\mathbf{C}_\psi(\delta_1 \otimes \delta_2) = \mathbf{C}_{1,\psi_1}(\delta_1) \otimes \mathbf{C}_{2,\psi_2}(\delta_2).$$

In particular, $\delta_k \in \Delta_k(\beta^{(k)})$, $k = 1,2$, if and only if $\delta_1 \otimes \delta_2 \in \Delta(\beta)$, and, in this case, $\mathbf{C}(\delta_1 \otimes \delta_2) = \mathbf{C}_1(\delta_1) \otimes \mathbf{C}_2(\delta_2)$.

Proof

(i) For every $\delta_1 \in \Delta_1$ and $\delta_2 \in \Delta_2$ we obtain

$$\begin{aligned}
\mathbf{I}(\delta_1 \otimes \delta_2) &= \int (a^{(1)} \otimes a^{(2)})(a^{(1)} \otimes a^{(2)})' d(\delta_1 \otimes \delta_2) \\
&= \int\int (a^{(1)} a^{(1)\prime}) \otimes (a^{(2)} a^{(2)\prime})\, d\delta_2\, d\delta_1 \\
&= (\textstyle\int a^{(1)} a^{(1)\prime} d\delta_1) \otimes (\textstyle\int a^{(2)} a^{(2)\prime} d\delta_2) \\
&= \mathbf{I}_1(\delta_1) \otimes \mathbf{I}_2(\delta_2),
\end{aligned}$$

by the usual rules for integration with respect to product measures (FUBINI's Theorem) and for doing calculus with KRONECKER products.

(ii) If ψ_k is identifiable under δ_k this implies $L_{k,\psi_k} = M_k \mathbf{I}_k(\delta_k)$ for a suitable matrix M_k. Hence, $L_\psi = (M_1 \mathbf{I}_1(\delta_1)) \otimes (M_2 \mathbf{I}_2(\delta_2)) = (M_1 \otimes M_2)(\mathbf{I}_1(\delta_1) \otimes \mathbf{I}_2(\delta_2))$ which yields the identifiability of ψ in view of (i). Therefore the covariance matrices involved exist and

$$\begin{aligned}
\mathbf{C}_\psi(\delta_1 \otimes \delta_2) &= L_\psi \mathbf{I}(\delta_1 \otimes \delta_2)^- L'_\psi \\
&= (L_{1,\psi_1} \otimes L_{2,\psi_2})(\mathbf{I}_1(\delta_1)^- \otimes \mathbf{I}_2(\delta_2)^-)(L'_{1,\psi_1} \otimes L'_{2,\psi_2}) \\
&= (L_{1,\psi_1} \mathbf{I}_1(\delta_1)^- L'_{1,\psi_1}) \otimes (L_{2,\psi_2} \mathbf{I}_2(\delta_2)^- L'_{2,\psi_2}) \\
&= \mathbf{C}_{1,\psi_1}(\delta_1) \otimes \mathbf{C}_{2,\psi_2}(\delta_2)\ .
\end{aligned}$$

In particular, the whole parameter vectors $\beta^{(1)}$, $\beta^{(2)}$, and β are identifiable if and only if the corresponding information matrices $\mathbf{I}_1(\delta_1)$, $\mathbf{I}_2(\delta_2)$, and $\mathbf{I}(\delta_1 \otimes \delta_2)$ are regular, i.e. if they have full rank. As $\text{rank}(\mathbf{I}(\delta_1 \otimes \delta_2)) = \text{rank}(\mathbf{I}_1(\delta_1))\text{rank}(\mathbf{I}_2(\delta_2))$ in view of (i) this proves the claimed equivalence. □

By Lemma 4.1 we get $\det(\mathbf{I}(\delta_1 \otimes \delta_2)) = \det(\mathbf{I}_1(\delta_1) \otimes \mathbf{I}_2(\delta_2)) = \det(\mathbf{I}_1(\delta_1))^{p_2} \det(\mathbf{I}_2(\delta_2))^{p_1}$ for every product design $\delta_1 \otimes \delta_2$. Hence, the best product design $\delta_1^* \otimes \delta_2^*$ (with respect to the D-criterion) is generated by the D-optimum designs δ_1^* and δ_2^* in the marginal models (cf BANDEMER et al. (1977), p 241).

The first result on the D-optimality of the best product design within the class of all designs was published by HOEL (1965) on polynomial regression and generalized by PETERSEN and KUKS (1971). With the present notation the proof simplifies substantially (cf RAFAJŁOWICZ and MYSZKA (1988)):

Theorem 4.2

Let δ_k^* be D-optimum in the single factor model $\mu_k(t_k) = a^{(k)}(t_k)'\beta^{(k)}$, $t_k \in T_k$. Then the product design $\delta^* = \delta_1^* \otimes \delta_2^*$ is D-optimum in the two-factor model with complete product-type interactions $\mu(t_1, t_2) = (a^{(1)}(t_1) \otimes a^{(2)}(t_2))'\beta$, $(t_1, t_2) \in T_1 \times T_2$.

Proof

The KIEFER-WOLFOWITZ equivalence theorem (Theorem 2.1) yields the inequality

$$a^{(k)}(t_k)'\mathbf{C}_k(\delta_k^*)a^{(k)}(t_k) \le p_k$$

for every $t_k \in T_k$ if δ_k^* is D-optimum in the corresponding single factor model. In view of Lemma 4.1 we obtain

$$a(t)'\mathbf{C}(\delta^*)a(t) = (a^{(1)}(t_1)'\mathbf{C}_1(\delta_1^*)a^{(1)}(t_1)) \otimes (a^{(2)}(t_2)'\mathbf{C}_2(\delta_2^*)a^{(2)}(t_2)) \le p_1 p_2 = p$$

for every $t = (t_1, t_2) \in T_1 \times T_2$. The KIEFER-WOLFOWITZ equivalence theorem, thus, proves the D-optimality of the product design $\delta^* = \delta_1^* \otimes \delta_2^*$. □

We turn to the A-criterion, next. Also for this criterion we obtain a straightforward factorization $\mathrm{tr}(\mathbf{C}(\delta_1 \otimes \delta_2)) = \mathrm{tr}(\mathbf{C}_1(\delta_1) \otimes \mathbf{C}_2(\delta_2)) = \mathrm{tr}(\mathbf{C}_1(\delta_1))\mathrm{tr}(\mathbf{C}_2(\delta_2))$ for every product design $\delta_1 \otimes \delta_2 \in \Delta(\beta)$ in view of Lemma 4.1. As in the preceding case of D-optimality the best product design $\delta_1^* \otimes \delta_2^*$ with respect to the A-criterion is generated by the A-optimum designs δ_1^* and δ_2^* from the marginal models. If we replace the KIEFER-WOLFOWITZ equivalence theorem in the proof of Theorem 4.2 by FEDOROV's equivalence theorem (Theorem 2.2) the A-optimality of the best product design can be extended to the whole class Δ of competing designs:

Theorem 4.3
Let δ_k^* be A-optimum in the single factor model $\mu_k(t_k) = a^{(k)}(t_k)'\beta^{(k)}$, $t_k \in T_k$. Then the product design $\delta^* = \delta_1^* \otimes \delta_2^*$ is A-optimum in the two-factor model with complete product-type interactions $\mu(t_1, t_2) = (a^{(1)}(t_1) \otimes a^{(2)}(t_2))'\beta$, $(t_1, t_2) \in T_1 \times T_2$.

Example 4.6 *Two-dimensional linear regression with interactions (cf Example 4.2 (i)):*
On a symmetric interval $T_k = [-\tau_k, \tau_k]$ the design δ_k^* which assigns equal weights $\frac{1}{2}$ to each of the endpoints $-\tau_k$ and τ_k of the interval is simultaneously D- and A-optimum in the marginal model of linear regression (see Example 3.9). Hence, by Theorem 4.2 resp. 4.3 the product design $\delta_1^* \otimes \delta_2^*$ is D- and A-optimum and it assigns equal weights $\frac{1}{4}$ to each of the four corner points $(\pm\tau_1, \pm\tau_2)$ of the rectangle $[-\tau_1, \tau_1] \times [-\tau_2, \tau_2]$. For the D-criterion the optimality of the equireplicated design concentrated on the corners carries over to arbitrary rectangular design regions $[\tau_{11}, \tau_{12}] \times [\tau_{21}, \tau_{22}]$ because of the invariance of the D-optimality with respect to linear transformations in linear regression. □

Another result of this type treats the c-optimality if the vector $c \in \mathbb{R}^p$ of interest can be factorized according to $c = c_1 \otimes c_2$ into components $c_1 \in \mathbb{R}^{p_1}$, $c_2 \in \mathbb{R}^{p_2}$ associated with the parameters of the marginal models. Because of the factorization of the variance $\mathbf{C}_{c'\beta}(\delta_1 \otimes \delta_2) = \mathbf{C}_{1,c_1'\beta^{(1)}}(\delta_1)\, \mathbf{C}_{2,c_2'\beta^{(2)}}(\delta_2)$ in view of Lemma 4.1 the product of the c_1- resp. c_2-optimum designs δ_1^* resp. δ_2^* in the marginal models is again the best product design with respect to the c-criterion ($c = c_1 \otimes c_2$). The extension to the c-optimality within the class Δ of all designs will be proved by means of ELFVING's theorem:

Theorem 4.4
Let $c = c_1 \otimes c_2$, $c_1 \in \mathbb{R}^{p_1}$, $c_2 \in \mathbb{R}^{p_2}$, and let δ_k^* be c_k-optimum in the single factor model $\mu_k(t_k) = a^{(k)}(t_k)'\beta^{(k)}$, $t_k \in T_k$. Then $\delta^* = \delta_1^* \otimes \delta_2^*$ is c-optimum in the two-factor model with complete product-type interactions $\mu(t_1, t_2) = (a^{(1)}(t_1) \otimes a^{(2)}(t_2))'\beta$, $(t_1, t_2) \in T_1 \times T_2$.

Proof
First we note that $\delta_k^* \in \Delta_k(c_k'\beta^{(k)})$, $k = 1, 2$, and hence $\delta_1^* \otimes \delta_2^* \in \Delta(c'\beta)$ in view of Lemma 4.1. ELFVING's theorem (Theorem 2.13) states that there is a c_k-optimum design $\tilde{\delta}_k = \sum_{n=1}^{N_k} w_{kn} \epsilon_{t_k^{(n)}}$ with finite support $\{t_k^{(1)}, ..., t_k^{(N_k)}\}$ and a positive constant γ_k such that the vector $\gamma_k c_k$ lies on the boundary of the convex hull of the marginal ELFVING set $\{a^{(k)}(t_k);\ t_k \in T_k\} \cup \{-a^{(k)}(t_k);\ t_k \in T_k\}$ and the stretched vector $\gamma_k c_k$ can be represented as

$$\gamma_k c_k = \sum_{n=1}^{N_k} z_{kn} w_{kn} a^{(k)}(t_k^{(n)})$$

for some sign variables $z_{kn} \in \{-1,1\}$, $n = 1, ..., N_k$.

Now the product design $\widetilde{\delta} = \widetilde{\delta}_1 \otimes \widetilde{\delta}_2 = \sum_{n_1=1}^{N_1} \sum_{n_2=1}^{N_2} w_{1n_1} w_{2n_2} \epsilon_{(t_1^{(n_1)}, t_2^{(n_2)})}$ is also supported on a finite set $\{(t_1^{(n_1)}, t_2^{(n_2)});\ n_1 = 1, ..., N_1,\ n_2 = 1, ..., N_2\}$. With $\gamma = \gamma_1\gamma_2 > 0$ we obtain

$$\gamma c = (\gamma_1 c_1) \otimes (\gamma_2 c_2) = \sum_{n_1=1}^{N_1} \sum_{n_2=2}^{N_2} z_{1n_1} z_{2n_2} w_{1n_1} w_{2n_2} a^{(1)}(t_1^{(n_1)}) \otimes a^{(2)}(t_2^{(n_2)})$$

where $z_{1n_1} z_{2n_2} \in \{-1,1\}$, $n_1 = 1, ..., N_1$, $n_2 = 1, ..., N_2$, and γc lies on the boundary of the convex hull of the ELFVING set $\{a^{(1)}(t_1) \otimes a^{(2)}(t_2);\ (t_1, t_2) \in T_1 \times T_2\} \cup \{-a^{(1)}(t_1) \otimes a^{(2)}(t_2);\ (t_1, t_2) \in T_1 \times T_2\}$. Applying ELFVING's theorem again we see that the product design $\widetilde{\delta} = \widetilde{\delta}_1 \otimes \widetilde{\delta}_2$ is c-optimum. As has been mentioned before $\delta^* = \delta_1^* \otimes \delta_2^*$ is the best product design and, hence, δ^* is c-optimum in $\Delta(c'\beta)$. □

In an obvious way the results of Theorems 4.2 to 4.4 can be extended to K factors, $K \geq 2$, by induction, e.g. if δ_k^* is c_k-optimum, $k = 1, ..., K$, then the K-fold product design $\delta^* = \bigotimes_{k=1}^K \delta_k^*$ is c-optimum for $c = \bigotimes_{k=1}^K c_k$. We summarize and expand these assertions in the following generalization of a theorem by RAFAJŁOWICZ and MYSZKA (1988) to potentially singular information matrices:

Theorem 4.5

A1: *ψ is an s-dimensional minimum linear aspect of β in the K-factor model $\mu(t_1, ..., t_K) = (\bigotimes_{k=1}^K a^{(k)}(t_k))'\beta$, $(t_1, ..., t_K) \in T_1 \times ... \times T_K$, for which the associated transformation matrix factorizes, $L_\psi = \bigotimes_{k=1}^K L_{k,\psi_k}$, and where ψ_k is an s_k-dimensional linear aspect of $\beta^{(k)}$ in the marginal model $\mu_k(t_k) = a^{(k)}(t_k)'\beta^{(k)}$, $t_k \in T_k$.*

A2: *Let $\varphi : \mathbb{R}^{s\times s} \to \mathbb{R}$ and $\varphi_k : \mathbb{R}^{s_k \times s_k} \to \mathbb{R}$. The criterion function $\Phi : \Delta \to \mathbb{R} \cup \{\infty\}$ is defined by $\Phi(\delta) = \varphi(\mathbf{C}_\psi(\delta))$ for $\delta \in \Delta(\psi)$ and $\Phi(\delta) = \infty$ otherwise, and the marginal criterion function $\Phi^{(k)} : \Delta_k \to \mathbb{R} \cup \{\infty\}$ is defined by $\Phi^{(k)}(\delta_k) = \varphi_k(\mathbf{C}_{k,\psi_k}(\delta_k))$ for $\delta_k \in \Delta_k(\psi_k)$ and $\Phi^{(k)}(\delta_k) = \infty$ otherwise.*

A3: *Φ and $\Phi^{(k)}$ are convex on Δ and Δ_k, respectively.*

A4: *For δ_k^* the gradient matrices ∇_{φ_k} at $\mathbf{C}_{k,\psi_k}(\delta_k^*)$ and ∇_φ at $\mathbf{C}_\psi(\bigotimes_{k=1}^K \delta_k^*)$ exist and the following factorization holds:*

$$\nabla_\varphi(\mathbf{C}_\psi(\bigotimes_{k=1}^K \delta_k^*)) = \bigotimes_{k=1}^K \nabla_{\varphi_k}(\mathbf{C}_{k,\psi_k}(\delta_k^*)) \ .$$

If the designs δ_k^* are $\Phi^{(k)}$-optimum in the marginal models and if the assumptions A1 to A4 hold, then the K-fold product design $\delta^* = \bigotimes_{k=1}^K \delta_k^*$ is Φ-optimum.

Proof

By induction we obtain from Lemma 4.1 that $\delta^* \in \Delta(\psi)$ and $\mathbf{C}_\psi(\delta^*) = \bigotimes_{k=1}^K \mathbf{C}_{k,\psi_k}(\delta_k^*)$. The equivalence theorem for possibly singular information matrices (Theorem 2.10) states that there exist matrices H_k such that $\text{rank}(H_k) = p_k - \varrho_k$, where $\varrho_k = \text{rank}(\mathbf{I}_k(\delta_k^*))$, $\mathbf{I}_k(\delta_k^*) + H_k'H_k$ is regular, and for every $t_k \in T_k$

$$\begin{aligned} & a^{(k)}(t_k)'(\mathbf{I}_k(\delta_k^*) + H_k'H_k)^{-1} L_{k,\psi_k}' \nabla_{\varphi_k}(\mathbf{C}_{k,\psi_k}(\delta_k^*)) L_{k,\psi_k} (\mathbf{I}_k(\delta_k^*) + H_k'H_k)^{-1} a^{(k)}(t_k) \\ \leq\ & \text{tr}(\mathbf{C}_{k,\psi_k}(\delta_k^*) \nabla_{\varphi_k}(\mathbf{C}_{k,\psi_k}(\delta_k^*))) \ . \end{aligned}$$

We can decompose the information matrices $\mathbf{I}_k(\delta_k^*)$ according to $\mathbf{I}_k(\delta_k^*) = F_k' F_k$ such that $F_k \in \mathbb{R}^{\varrho_k \times p_k}$ and $\operatorname{rank}(F_k) = \varrho_k$. Define $G_1 = H_1$ and

$$G_k = \begin{pmatrix} G_{k-1} \otimes H_k \\ G_{k-1} \otimes F_k \\ (\bigotimes_{\kappa=1}^{k-1} F_\kappa) \otimes H_k \end{pmatrix},$$

$k \geq 2$. Then the matrix G_k has $\prod_{\kappa=1}^{k} p_\kappa$ columns,

$$\begin{aligned} \operatorname{rank}(G_k) &\leq \operatorname{rank}(G_{k-1} \otimes H_k) + \operatorname{rank}(G_{k-1} \otimes F_k) + \operatorname{rank}((\textstyle\bigotimes_{\kappa=1}^{k-1} F_\kappa) \otimes H_k) \\ &= p_k \operatorname{rank}(G_{k-1}) + (p_k - \varrho_k) \textstyle\prod_{\kappa=1}^{k-1} \varrho_\kappa \end{aligned}$$

and

$$\textstyle\bigotimes_{\kappa=1}^{k} \mathbf{I}_\kappa(\delta_\kappa^*) + G_k' G_k = (\bigotimes_{\kappa=1}^{k-1} \mathbf{I}_\kappa(\delta_\kappa^*) + G_{k-1}' G_{k-1}) \otimes (\mathbf{I}_k(\delta_k^*) + H_k' H_k) .$$

By induction we obtain $\operatorname{rank}(G_k) \leq \prod_{\kappa=1}^{k} p_\kappa - \prod_{\kappa=1}^{k} \varrho_\kappa$ and $\bigotimes_{\kappa=1}^{k} \mathbf{I}_\kappa(\delta_\kappa^*) + G_k' G_k = \bigotimes_{\kappa=1}^{k} (\mathbf{I}_\kappa(\delta_\kappa^*) + H_\kappa' H_\kappa)$ and, In particular, for $k = K$ we obtain $\operatorname{rank}(G_K) \leq p - \varrho$ where $\varrho = \operatorname{rank}(\mathbf{I}(\delta^*))$ and $\mathbf{I}(\delta^*) + G_K' G_K = \bigotimes_{k=1}^{K} (\mathbf{I}_k(\delta_k^*) + H_k' H_k)$. Hence $\mathbf{I}(\delta^*) + G_K' G_K$ is regular and $\operatorname{rank}(G_K) = p - \varrho$. Furthermore, all vectors and matrices factorize and

$$\begin{aligned} &a(t)'(\mathbf{I}(\delta^*) + G_K' G_K)^{-1} L_\psi' \nabla_\varphi(\mathbf{C}_\psi(\delta^*)) L_\psi (\mathbf{I}(\delta^*) + G_K' G_K)^{-1} a(t) \\ &= \textstyle\bigotimes_{k=1}^{K} (a^{(k)}(t_k)'(\mathbf{I}_k(\delta_k^*) + H_k' H_k)^{-1} L_{k,\psi_k}' \nabla_{\varphi_k}(\mathbf{C}_{k,\psi_k}(\delta_k^*)) L_{k,\psi_k} (\mathbf{I}_k(\delta_k^*) + H_k' H_k)^{-1} a^{(k)}(t_k)) \\ &\leq \textstyle\prod_{k=1}^{K} \operatorname{tr}(\mathbf{C}_{k,\psi_k}(\delta_k^*) \nabla_{\varphi_k}(\mathbf{C}_{k,\psi_k}(\delta_k^*))) \\ &= \operatorname{tr}(\mathbf{C}_\psi(\delta^*) \nabla_\varphi(\mathbf{C}_\psi(\delta^*))) , \end{aligned}$$

for every $t = (t_1, ..., t_k) \in T_1 \times ... \times T_K$. The result follows by an application of Theorem 2.10 again. □

From this result we can recover Theorems 4.2 and 4.3 and their generalization to the Φ_q-criteria, for $q < \infty$, by consideration of the full parameter vectors $\psi_k(\beta^{(k)}) = \beta^{(k)}$ and Theorem 4.4 by consideration of $\psi_k(\beta^{(k)}) = c_k' \beta^{(k)}$. For more general linear aspects ψ the D_ψ- and A_ψ-optimality can be obtained from Theorem 4.5 as exhibited next:

Corollary 4.6

Let ψ, ψ_k be linear aspects in the K-factor model (4.2) and the corresponding marginal models (4.1), respectively, with $L_\psi = \bigotimes_{k=1}^{K} L_{k,\psi_k}$.

(i) *Let ψ_k be minimum. If δ_k^* is D_{ψ_k}-optimum in the kth marginal model, then the K-fold product design $\delta^* = \bigotimes_{k=1}^{K} \delta_k^*$ is D_ψ-optimum in the K-factor model with complete product-type interactions.*

(ii) *If δ_k^* is A_{ψ_k}-optimum in the kth marginal model, then the K-fold product design $\delta^* = \bigotimes_{k=1}^{K} \delta_k^*$ is A_ψ-optimum in the K-factor model with complete product-type interactions.*

Proof

(i) The minimality of ψ_k ensures that L_{ψ_k} has full row rank. Hence L_ψ has full row rank and ψ is minimum. Now choosing φ and φ_k as $-\ln\det$ for matrices of the appropriate dimensions we can directly apply Theorem 4.5.

(ii) There are minimum linear aspects $\widetilde{\psi}_k$ with $L'_{\widetilde{\psi}_k} L_{\widetilde{\psi}_k} = L'_{\psi_k} L_{\psi_k}$ and, consequently, $\mathrm{tr}(\mathbf{C}_{k,\psi_k}(\delta_k)) = \mathrm{tr}(\mathbf{C}_{k,\widetilde{\psi}_k}(\delta_k))$ for $\delta_k \in \Delta_k(\psi_k) = \Delta_k(\widetilde{\psi}_k)$. The linear aspect $\widetilde{\psi} = \bigotimes_{k=1}^K \widetilde{\psi}_k$ is also minimum with $L'_{\widetilde{\psi}} L_{\widetilde{\psi}} = L'_\psi L_\psi$, such that $\mathrm{tr}(\mathbf{C}_\psi(\delta)) = \mathrm{tr}(\mathbf{C}_{\widetilde{\psi}}(\delta))$ for $\delta \in \Delta(\psi) = \Delta(\widetilde{\psi})$. Thus the A_ψ- and $A_{\widetilde{\psi}}$-criterion resp. the A_{ψ_k}- and $A_{\widetilde{\psi}_k}$-criterion, coincide. By letting φ and φ_k equal to the trace we obtain the $A_{\widetilde{\psi}}$- and, hence, thr A_ψ-optimality of the product design $\delta^* = \bigotimes_{k=1}^K \delta_k^*$ by Theorem 4.5. □

Example 4.7 *Quadratic regression (cf Example 4.2 (ii)):*

For the whole parameter vector $\beta^{(k)}$ the D-optimum design δ_k^* assigns equal weights $\frac{1}{3}$ to the endpoints -1 and 1 and to the midpoint 0 of the interval (cf Example 3.9). By applying Theorem 4.2 (or Corollary 4.6) we recover that the product design $\delta^* = \delta_1^* \otimes \delta_2^*$ which assigns equal weights $\frac{1}{9}$ to the four corner points $(-1,-1), (-1,1), (1,-1)$, and $(1,1)$, to the four midpoints of the edges $(-1,0), (0,-1), (0,1)$, and $(1,0)$ and to the center point $(0,0)$ of the square is D-optimum.

As the D-optimality is invariant with respect to translations the equireplicated design supported on the corners, on the midpoints of the edges and on the center is also D-optimum for every rectangular design region $T = [\tau_{11}, \tau_{12}] \times [\tau_{21}, \tau_{22}]$, i.e. $\delta^* = \frac{1}{9} \sum_{i=0}^2 \sum_{j=0}^2 \epsilon_{(\tau_{1i}, \tau_{2j})}$ where $\tau_{k0} = \frac{1}{2}(\tau_{k1} + \tau_{k2})$.

If we are interested in the lower order terms without quadratic effects, the relevant linear aspect are given by $\psi_k(\beta^{(k)}) = (\beta_1^{(k)}, \beta_2^{(k)})'$ and $\psi(\beta) = (\beta_0, \beta_1, \beta_2, \beta_{12})'$. It can be checked (cf PUKELSHEIM (1980)) that the D_{ψ_k}-optimum design δ_k^* assigns weights $\frac{1}{4}$ to each of the endpoints and $\frac{1}{2}$ to the midpoint of the interval. By Corollary 4.6 the product design $\delta^* = \delta_1^* \otimes \delta_2^*$ which assigns weights $\frac{1}{16}$ to the corners, $\frac{1}{8}$ to the midpoint of the edges and $\frac{1}{4}$ to the center of the square is D_ψ-optimum. (Note that this criterion is not invariant with respect to translations.) □

Corollary 4.6 can be applied directly to the problem of choosing an optimum design to minimize the weighted average of the variance of the prediction $\int a(t)'\mathbf{C}(\delta)a(t)\xi(dt)$ if the weighting measure also factorizes, $\xi = \bigotimes_{k=1}^K \xi_k$,

Corollary 4.7

Let ξ_k be weighting measures on T_k and $\xi = \bigotimes_{k=1}^K \xi_k$. If δ_k^* is Q-optimum with respect to ξ_k in the kth marginal model, then the K-fold product design $\delta^* = \bigotimes_{k=1}^K \delta_k^*$ is Q-optimum with respect to ξ in the K-factor model with complete product-type interactions.

Proof

We can decompose $\int a^{(k)}a^{(k)\prime}d\xi_k = L_k'L_k$ such that L_k can be regarded as the matrix associated with a linear aspect ψ_k and the criterion function of the Q-criterion coincides with that of the A_{ψ_k}-criterion,

$$\int a^{(k)\prime}\mathbf{C}_k(\delta_k)a^{(k)}d\xi_k = \mathrm{tr}(\int a^{(k)}a^{(k)\prime}d\xi_k\, \mathbf{C}_k(\delta_k)) = \mathrm{tr}(L_k\mathbf{C}_k(\delta_k)L_k') = \mathrm{tr}(\mathbf{C}_{k,\psi_k}(\delta_k))\ .$$

Similarly, $\bigotimes_{k=1}^K L_k$ can be regarded as the matrix associated with a linear aspect ψ in the K-factor model and

$$\int a'\mathbf{C}(\delta)a\, d\xi = \mathrm{tr}((\bigotimes_{k=1}^K \int a^{(k)}a^{(k)\prime}d\xi_k)\mathbf{C}(\delta)) = \mathrm{tr}(\mathbf{C}_\psi(\delta))\ .$$

With this interpretation of the Q-criterion the result follows from Corollary 4.6. □

We now turn to the situation where we are interested in the effects of a single factor and where the other components of the parameter vector may be considered as systematic noise, e. g. as block effects. In this case we may assume that all these side effects are subsumed in the second factor and that there is a constant term present in the second marginal model, at least implicitly: $\mathbf{1} \in \operatorname{span}(a_1^{(2)}, ..., a_{p_2}^{(2)})$, i. e. there is a $\ell \in \mathbb{R}^{p_2}$ such that $\sum_{i=1}^{p_2} \ell_i a_i^{(2)}(t_2) = 1$ for all $t_2 \in T_2$. Without loss of generality we can consider a parametrization of the model in which the effect of the first factor, i. e. the factor of interest, is explicitly separated from the side effects

$$\mu(t_1, t_2) = a^{(1)}(t_1)'\beta_1 + (a^{(1)}(t_1) \otimes f_2(t_2))'\beta_{12} , \tag{4.4}$$

$(t_1, t_2) \in T_1 \times T_2$. The parameters of interest are collected in β_1. We can identify (4.4) as a special case of the general two-factor model (4.3) with $a^{(2)} = (\mathbf{1}, f_2')'$. The corresponding marginal models are given by the usual model (4.1) for the first factor, $\mu_1(t_1) = a^{(1)}(t_1)'\beta_1$, $t_1 \in T_1$, and by a response function

$$\mu_2(t_2) = \beta_0 + f_2(t_2)'\beta_1^{(2)} , \tag{4.5}$$

$t_2 \in T_2$, for the second factor where an explicit constant (or intercept) term is present. If we look for an A- or D-optimum design for the whole parameter vector β_1 associated with the effects of the first factor, or for some linear aspect of β_1, then we can obtain the following result as a straightforward consequence of Corollary 4.6,

Corollary 4.8

Let ψ be a linear aspect in the two-factor model $\mu_1(t_1, t_2) = a^{(1)}(t_1)'\beta_1 + (a^{(1)}(t_1) \otimes f_2(t_2))'\beta_{12}$, $(t_1, t_2) \in T_1 \times T_2$, which depends on β_1, $\psi(\beta) = \psi_1(\beta_1)$.

(i) *Let ψ_1 be minimum. If δ_1^* is D_{ψ_1}-optimum in the first marginal model $\mu_1(t_1) = a^{(1)}(t_1)'\beta^{(1)}$, $t_1 \in T_1$, and if δ_2^* is β_0-optimum in the second marginal model $\mu_2(t_2) = \beta_0 + f_2(t_2)'\beta_1^{(2)}$, $t_2 \in T_2$, then the product design $\delta^* = \delta_1^* \otimes \delta_2^*$ is D_ψ-optimum in the two-factor model.*

(ii) *If δ_1^* is A_{ψ_1}-optimum in the first marginal model $\mu_1(t_1) = a^{(1)}(t_1)'\beta^{(1)}$, $t_1 \in T_1$, and if δ_2^* is β_0-optimum in the second marginal model $\mu_2(t_2) = \beta_0 + f_2(t_2)'\beta_1^{(2)}$, $t_2 \in T_2$, then the product design $\delta^* = \delta_1^* \otimes \delta_2^*$ is A_ψ-optimum in the two-factor model.*

Example 4.8 *Intra-class model with identical partial models (cf Example 4.3):*

If we assume that a constant term is explicitly involved in f the intraclass mode can be written as $\mu(i, u) = \alpha_i + f_2(u)'\beta_{2,i}$. If we are interested in the direct effects $\psi(\beta) = (\alpha_i)_{i=1,...,I}$, then the product design $\delta^* = \delta_1^* \otimes \delta_2^*$ is D- (and A-)optimum where δ_1^* assigns equal weights $\frac{1}{I}$ to each level $i = 1, ..., I$ of the first factor of interest (see Example 3.3) and δ_2^* is the optimum design for β_0 in the marginal model $\mu_2(u) = \beta_0 + f_2(u)'\beta_2$, $u \in U$. In particular, in case of linear regression in the second factor the β_0-optimum design can be taken either as the degenerate one-point design at 0, $\delta_2^* = \epsilon_0$, if $0 \in U$, or the optimum extrapolation design of Example 2.1 has to be used. □

In general, in the situation that the evaluation of the constant term β_0 can be achieved by an interpolation within the second marginal model we obtain a general complete class theorem which states that for any design there is a product design which performs at least as good as δ in the sense that the covariance matrix is not increased for every linear aspect ψ_1 of β_1,

Theorem 4.9

Let ψ be a linear aspect in the two-factor model $\mu_1(t_1,t_2) = a^{(1)}(t_1)'\beta_1 + (a^{(1)}(t_1) \otimes f_2(t_2))'\beta_{12}$, $(t_1,t_2) \in T_1 \times T_2$, which depends on β_1, $\psi(\beta) = \psi_1(\beta_1)$.

If $\mathbf{0}$ lies in the convex hull of $\{f_2(t_2);\, t_2 \in T_2\}$ and if δ_2^* is β_0-optimum in the second marginal model $\mu_2(t_2) = \beta_0 + f_2(t_2)'\beta_1^{(2)}$, $t_2 \in T_2$, then $\mathbf{C}_\psi(\delta) \geq \mathbf{C}_\psi(\delta_1 \otimes \delta_2^*)$ for every $\delta \in \Delta(\psi)$, where δ_1 is the first marginal of δ.

Proof

First we notice that $\delta \in \Delta(\psi)$ implies $\delta_1 \in \Delta_1(\psi_1)$ by Lemma 3.5 because the first marginal model can be considered as a submodel of the two-factor model. Hence by Lemma 4.1 we get $\delta_1 \otimes \delta_2^* \in \Delta(\psi)$. Now $\mathbf{C}_{2,\beta_0}(\delta_2^*) = 1$ by Corollary 2.14 and it follows that $\mathbf{C}_\psi(\delta_1 \otimes \delta_2^*) = \mathbf{C}_{1,\psi_1}(\delta_1)$ by Lemma 4.1. Hence, the refinement argument of Lemma 3.5 completes the proof. □

For the whole marginal parameter vector β_1 associated with the first factor we obtain $\mathbf{C}_{\beta_1}(\delta) \geq \mathbf{C}_{\beta_1}(\delta_1 \otimes \delta_2^*)$ for $\delta \in \Delta(\beta_1)$. It should be added that in Theorem 4.9 the assumption on $\mathbf{0}$ lying in the convex hull of $\{f_2(t_2);\, t_2 \in T_2\}$ can be removed. In case $\delta_2^* \in \Delta_2(\beta_2)$ this can be done by making use of an equivalence theorem due to HUANG and HSU (1993) for the situation of a given marginal design for the first factor. The general case can be treated after a preceding application of an additional refinement argument.

To complete our survey on optimality of product designs for models with complete product-type interactions we mention a result on E-optimality for which the criterion function does not necessarily satisfy the differentiability conditions of Theorem 4.5,

Theorem 4.10

Let δ_k^* be E-optimum in the single factor model $\mu_k(t_k) = a^{(k)}(t_k)'\beta^{(k)}$, $t_k \in T_k$. Then the product design $\delta^* = \delta_1^* \otimes \delta_2^*$ is E-optimum in the two-factor model with complete product-type interactions $\mu(t_1,t_2) = (a^{(1)}(t_1) \otimes a^{(2)}(t_2))'\beta$, $(t_1,t_2) \in T_1 \times T_2$.

Proof

For an E-optimum design δ_k^* there exists a positive semidefinite $p_k \times p_k$ matrix M_k such that $\mathrm{tr}(M_k) = 1$ and $a^{(k)}(t_k)'M_k a^{(k)}(t_k) \leq \lambda_{\max}(\mathbf{C}_k(\delta_k^*))^{-1}$ for every $t_k \in T_k$ by the equivalence Theorem 2.12. Let $M = M_1 \otimes M_2$. Then M is also positive semidefinite, $\mathrm{tr}(M) = \mathrm{tr}(M_1)\,\mathrm{tr}(M_2) = 1$, and

$$\begin{aligned} a(t)'Ma(t) &= (a^{(1)}(t_1)'M_1a^{(1)}(t_1)) \otimes (a^{(2)}(t_2)'M_2a^{(2)}(t_2)) \\ &\leq \lambda_{\max}(\mathbf{C}_1(\delta_1^*))^{-1}\lambda_{\max}(\mathbf{C}_2(\delta_2^*))^{-1} \\ &= \lambda_{\max}(\mathbf{C}(\delta_1^* \otimes \delta_2^*))^{-1}\,. \end{aligned}$$

Again by Theorem 2.12 the product design $\delta_1^* \otimes \delta_2^*$ is E-optimum. □

Example 4.9 *Two-dimensional linear regression with interactions (cf Example 4.2 (i)):*

On a symmetric interval $T_k = [-\tau_k, \tau_k]$ the design δ_k^* which assigns equal weights $\frac{1}{2}$ to each of the endpoints $-\tau_k$ and τ_k of the interval is also E-optimum in the marginal model of linear regression (see Example 3.9). Hence, by Theorem 4.10 the product design $\delta_1^* \otimes \delta_2^*$ which assigns equal weights $\frac{1}{4}$ to each of the four corner points $(\pm\tau_1, \pm\tau_2)$ of the rectangle $[-\tau_1, \tau_1] \times [-\tau_2, \tau_2]$ is E-optimum. (For the simultaneous D- and A-optimality of the same design we recall Example 4.6.) □

Concluding remark. In models with complete product-type interaction optimum design can be generated as products of those marginals which are optimum in the corresponding marginal models. The first result of this kind was obtained by HOEL (1965) and extended by PETERSEN and KUKS (1971) to general regression functions. Φ_q-criteria were treated by RAFAJŁOWICZ and MYSZKA (1988) and optimality in case of a given marginal design was obtained by COOK and THIBODEAU (1980) and HUANG and HSU (1993) by using a conditional approach.

5 No Interactions

In many practically relevant experimental situations no interactions occur between the factors of influence. Also in these cases there are many design problems in which product designs are optimum and, again, the associated marginal designs are optimum in the corresponding marginal models. In particular, D-optimality is included if there is a constant term in the model under consideration. Such additive models with an explicit constant term will be treated in the first subsection. The second subsection deals with the case that for all or, at least, for all but one factors the regression functions are centered with respect to the optimum designs in the marginal models and the product of them results in orthogonal estimators of the parameters associated with the effects of the single factors in the whole model.

5.1 Additive Models

In this subsection we consider the general case of a model including explicitly a constant (or intercept) term, i. e. one of the regression functions is constant equal to 1 on the design region. Such K-factor models without interactions are called *additive models* (cf COOK and THIBODEAU, 1980). They are defined by

$$\mu(t_1,...,t_K) = a(t_1,...,t_K)'\beta = \beta_0 + \sum_{k=1}^{K} f_k(t_k)'\beta_k \ , \tag{5.1}$$

$t = (t_1,...,t_K) \in T = T_1 \times ... \times T_K$, such that $a(t_1,...,t_K) = (1, f_1(t_1)',...,f_K(t_K)')'$, $\beta_k \in \mathbb{R}^{p_k-1}$, and $\beta = (\beta_0, \beta_1',...,\beta_K')' \in \mathbb{R}^p$ with $p = \sum_{k=1}^{K} p_k - K + 1$. The corresponding marginal models also include a constant term each:

$$\mu_k(t_k) = a^{(k)}(t_k)'\beta^{(k)} = \beta_0 + f_k(t_k)'\beta_k \ , \tag{5.2}$$

$t_k \in T_k$, such that $a^{(k)}(t_k) = (1, f_k(t_k)')'$, $\beta^{(k)} = (\beta_0, \beta_k')' \in \mathbb{R}^{p_k}$, $k = 1,...,K$. Again, as in the preceding section we denote by sub- und superscripts k and (k) designs, parameters, aspects, etc. associated with the kth marginal model. We will be mainly interested in the two-factor model without interactions

$$\mu(t_1,t_2) = \beta_0 + f_1(t_1)'\beta_1 + f_2(t_2)'\beta_2 \ , \tag{5.3}$$

$(t_1,t_2) \in T_1 \times T_2$, such that $a(t_1,t_2) = (1, f_1(t_1)', f_2(t_2)')'$ and $\beta = (\beta_0, \beta_1', \beta_2')' \in \mathbb{R}^p$ with $p = p_1 + p_2 - 1$. The corresponding results for K factors can be derived by subsuming single factors into a higher dimensional meta-factor (see Corollary 5.3 and Theorem 5.14).

The main issue of this subsection is to show that marginal designs δ_k which are *good* for β_k in the kth marginal model generate a product design $\delta_1 \otimes \delta_2$ which is *good* for β_1 and β_2 in the whole additive model. For the D-criterion this carries over to the whole parameter vectors $\beta^{(1)}, \beta^{(2)}$ and β, respectively (see Theorem 5.2). Furthermore, as the D-criterion is not affected by reparametrizations the explicit appearance of the constant term in (5.1) to (5.3) is immaterial and can be substituted by a more general condition as will be indicated at the end of this subsection (Corollary 5.10 and Corollary 5.18, cf also the comment preceding formula (4.4)).

For illustrative purposes we present some examples which give an impression of the variety of models covered by (5.3),

Example 5.1 *Two-way layout without interactions:*
In this example we have two qualitative factors, and the joint effect of a level combination is the sum of the effects associated with the single factors. Hence, this model is given by

$$\mu(i,j) = \mu_0 + \alpha_i^{(1)} + \alpha_j^{(2)} ,$$

$i = 1, ..., I$, $j = 1, ..., J$. As there are only $I + J - 1$ linearly independent regression functions on $T = \{1, ..., I\} \times \{1, ..., J\}$ which constitute the response function this model is overparametrized. Therefore, we have to reduce the $I + J + 1$ ambigious parameters $\mu_0, \alpha_1^{(1)}, ..., \alpha_I^{(1)}, \alpha_1^{(2)}, ..., \alpha_J^{(2)}$ by the appropriate number $(I + J + 1) - (I + J - 1) = 2$ of identifiability conditions to arrive at a minimum parametrization (see Example 3.7). A natural choice with regard to the model under consideration is the parametrization of complete control, i. e. I and J are control levels for the first and second factor, respectively, and the identifiability conditions $\alpha_I^{(1)} = 0$ and $\alpha_J^{(2)} = 0$ are imposed. With these conditions the parameters have the following interpretation:

$\mu_0 = \mu(I, J)$ is the response on the control level combination (I, J);

$\alpha_i^{(1)} = \frac{1}{J}\sum_{j=1}^{J}(\mu(i,j) - \mu(I,j)) = \mu(i,j) - \mu(I,j)$, for every $j = 1, ..., J$, is the effect of level i for the first factor compared to control, $i = 1, ..., I - 1$;

and, analogously,

$\alpha_j^{(2)} = \frac{1}{I}\sum_{i=1}^{I}(\mu(i,j) - \mu(i,J)) = \mu(i,j) - \mu(i,J)$, for every $i = 1, ..., I$, is the effect of level j for the second factor compared to control, $j = 1, ..., J - 1$.

The vector of unknown parameters is thus $\beta = (\mu_0, \alpha_1^{(1)}, ..., \alpha_{I-1}^{(1)}, \alpha_1^{(2)}, ..., \alpha_{J-1}^{(2)})' \in \mathbb{R}^{I-J+1}$ and the regression functions are given by $a(i,j) = (1, \mathbf{e}'_{I-1,i}, \mathbf{e}'_{J-1,j})'$. We recall that e. g. $\mathbf{e}_{I-1,i}$ is a vector of length $I - 1$ with the ith entry equal to one and all other entries equal to zero; in particular $\mathbf{e}_{I-1,I}$ denotes a vector with all entries equal to zero ($\mathbf{e}_{I-1,I} = \mathbf{0}$).

Hence, with $f_1(i) = \mathbf{e}_{I-1,i}$ and $f_2(j) = \mathbf{e}_{J-1,j}$ we have one-way layouts with control (see Example 3.4), $\mu_1(i) = \mu_0 + \alpha_i^{(1)}$, $i = 1, ..., I$, with identifiability condition $\alpha_I^{(1)} = 0$, and $\mu_2(j) = \mu_0 + \alpha_j^{(2)}$, $j = 1, ..., J$, with $\alpha_J^{(2)} = 0$, respectively, as the corresponding marginal models for each of both factors.

By δ_1 and δ_2 we denote the marginals of a design δ on $T = \{1, ..., I\} \times \{1, ..., J\}$, i. e. $\delta_1(\{i\}) = \sum_{j=1}^{J} \delta(\{(i,j)\})$ and $\delta_2(\{j\}) = \sum_{i=1}^{I} \delta(\{(i,j)\})$. In particular, δ_1 and δ_2 are marginal designs on the marginal design regions $\{1, ..., I\}$ and $\{1, ..., J\}$, respectively. Let $\mathbf{w}_1 = (\delta_1(\{i\}))_{i=1,...,I-1}$ and $\mathbf{w}_2 = (\delta_2(\{j\}))_{j=1,...,J-1}$ be the vectors of the marginal weights for the treatment levels and let $\mathbf{I}_{12} = (\delta(\{(i,j)\}))_{i=1,...,I-1}^{j=1,...,J-1}$ be the matrix of weights of the treatment level combinations for the two factors. Then the information matrix of δ can be written as

$$\mathbf{I}(\delta) = \begin{pmatrix} 1 & \mathbf{w}_1' & \mathbf{w}_2' \\ \mathbf{w}_1 & \operatorname{diag}(\mathbf{w}_1) & \mathbf{I}_{12} \\ \mathbf{w}_2 & \mathbf{I}_{12}' & \operatorname{diag}(\mathbf{w}_2) \end{pmatrix}$$

where $\operatorname{diag}(\mathbf{w}_k)$ stands for a diagonal matrix with the entries of the vector $\mathbf{w}_k$ as its diagonal elements. Moreover, we obtain

$$\mathbf{I}_k(\delta_k) = \begin{pmatrix} 1 & \mathbf{w}_k' \\ \mathbf{w}_k & \operatorname{diag}(\mathbf{w}_k) \end{pmatrix}.$$

for the kth marginal information matrix. □

Example 5.2 *Polynomial regression without composite terms:*
In this model with two quantitative factors

$$\mu(t_1,t_2) = \beta_0 + \sum_{i=1}^{p_1-1} t_1^i \beta_{1i} + \sum_{j=1}^{p_2-1} t_2^j \beta_{2j} ,$$

$t = (t_1, t_2) \in T = T_1 \times T_2$, the marginal regression functions are the monomials in the single factors $f_k(t_k) = (t_k^i)_{i=1,\ldots,p_k-1}$, and each marginal model $\mu_k(t_k) = \beta_0 + \sum_{i=1}^{p_k-1} t_k^i \beta_{ki}$, $t_k \in T_k$, is a polynomial regression of degree $p_k - 1$ (see Example 1.2).

For linear regression, i. e. $p_1 = p_2 = 2$, we have the two-dimensional linear regression model

(i) $$\mu(t_1,t_2) = \beta_0 + \beta_1 t_1 + \beta_2 t_2 .$$

For quadratic marginals ($p_1 = p_2 = 3$) we consider the two-dimensional quadratic regression

(ii) $$\mu(t_1,t_2) = \beta_0 + \beta_{11} t_1 + \beta_{12} t_1^2 + \beta_{21} t_2 + \beta_{22} t_2^2$$

without an interaction term (compare this model with the two other two-dimensional quadratic regression models considered in Examples 4.2 and 7.4). □

Example 5.3 *One-way layout with additional regression:*
In this model with two different kinds of factors,

$$\mu(i,u) = \mu_0 + \alpha_i + f(u)'\beta_2 ,$$

$t = (i, u) \in \{1, \ldots, I\} \times U$, we assume that there is a control level for the first, qualitative factor which is described by the identifiability condition $\alpha_I = 0$ (cf Example 5.1). Then the first marginal model is a one-way layout with control, $\mu_1(i) = \mu_0 + \alpha_i$, $i = 1, \ldots, I$, ($\alpha_I = 0$; cf Example 3.4). For the second factor we have a regression model $\mu_2(u) = \beta_0 + f(u)'\beta_2$, $u \in U$, involving a constant term. For the interpretation of μ_0 and α_i in the additive model we refer to Example 5.1. The *regression* part will typically be described by a polynomial regression function, $f(u) = (u, u^2, \ldots, u^{p_2-1})'$.

The more classical case of a standard parametrization (without general mean) in the one-way layout part,

$$\mu(i,u) = \tilde{\alpha}_i + f(u)'\beta_2 ,$$

will be treated in the second subsection for some special f (see Example 5.14) and, in greater generality, in Section 6 (Corollary 6.15). We note again that this reparametrization does not affect the D-optimality. □

We start with the following essentially complete class result which states that every design δ is dominated by the product $\delta_1 \otimes \delta_2$ of its marginals δ_1 and δ_2 with respect to the D-criterion,

Lemma 5.1
In an additive model we have

$$\det(\mathbf{I}(\delta)) \le \det(\mathbf{I}_1(\delta_1)) \det(\mathbf{I}_2(\delta_2)) = \det(\mathbf{I}(\delta_1 \otimes \delta_2))$$

for every design $\delta \in \Delta$ with marginals δ_1 and δ_2, where $\mathbf{I}_k(\delta_k) = \int a^{(k)} a^{(k)\prime}\, d\delta_k$ is the information matrix in the kth marginal model.

Proof

Let $f(t_1,t_2) = (f_1(t_1)', f_2(t_2)')'$ and $\widetilde{f}_k = f_k - \int f_k \, d\delta_k$ be the centered versions of f_k, $k = 1, 2$, then $\widetilde{f}$ with $\widetilde{f}(t_1,t_2) = (\widetilde{f}_1(t_1)', \widetilde{f}_2(t_2)')'$ is the centered version of f (see Subsection 3.1). By Lemma 3.2 and the formula for the determinant of a partitioned matrix (Lemma A.2) we obtain

$$\begin{aligned} \det(\mathbf{I}(\delta)) &= \det(\textstyle\int \widetilde{f}\widetilde{f}' \, d\delta) \\ &\leq \det(\textstyle\int \widetilde{f}_1\widetilde{f}_1' \, d\delta_1) \det(\textstyle\int \widetilde{f}_2\widetilde{f}_2' \, d\delta_2) \\ &= \det(\mathbf{I}_1(\delta_1)) \det(\mathbf{I}_2(\delta_2)) \end{aligned}$$

with equality for $\delta = \delta_1 \otimes \delta_2$ as $\int \widetilde{f}_1\widetilde{f}_2' \, d(\delta_1 \otimes \delta_2) = \int \widetilde{f}_1 \, d\delta_1 \int \widetilde{f}_2' \, d\delta_2 = \mathbf{0}$. □

With respect to the D-criterion we can now give a complete solution of the design problem in additive models which shows that the product of the D-optimum designs in the marginal models forms a D-optimum design in the whole model without interactions.

Theorem 5.2

Let δ_k^* be D-optimum in the single factor model $\mu_k(t_k) = \beta_0 + f_k(t_k)'\beta_k$, $t_k \in T_k$. Then the product design $\delta^* = \delta_1^* \otimes \delta_2^*$ is D-optimum in the additive two-factor model $\mu(t_1,t_2) = \beta_0 + f_1(t_1)'\beta_1 + f_2(t_2)'\beta_2$, $(t_1,t_2) \in T_1 \times T_2$.

Proof

By Lemma 5.1 the determinant of the information matrix for a product design factorizes according to $\det(\mathbf{I}(\delta_1 \otimes \delta_2)) = \det(\mathbf{I}_1(\delta_1)) \det(\mathbf{I}_2(\delta_2))$ into the determinants of the information matrices in the marginal models. Hence D-optimum δ_1^* and δ_2^* yield the best product design $\delta_1^* \otimes \delta_2^*$ with respect to the D-criterion. Now, by the complete class result of Lemma 5.1 the product design $\delta_1^* \otimes \delta_2^*$ is also D-optimum within the class Δ of all designs. □

Whether $\delta^* = \delta_1^* \otimes \delta_2^*$ is the unique D-optimum design depends on the structure of the underlying regression functions, even if the marginal designs δ_1^* and δ_2^* are the unique D-optimum designs in the marginal models. The next example exhibits a situation where the resulting D-optimum design is, indeed, unique,

Example 5.4 *Two-way layout without interactions (complete control; cf Example 5.1):*

In the marginal one-way layout models the designs δ_1^* and δ_2^* which assign equal weights $\frac{1}{I}$ and $\frac{1}{J}$, respectively, to each level are D-optimum (see Example 3.3). Hence by Theorem 5.2 the product design $\delta^* = \delta_1^* \otimes \delta_2^*$ which assigns equal weights $\frac{1}{IJ}$ to each of the level combinations (i,j) is D-optimum in the two-way layout without interactions. Note that the same design is also D-optimum for the two-way layout model with complete interactions (see Example 4.1).

Recall that for every design δ we have

$$\mathbf{I}(\delta) = \begin{pmatrix} 1 & \mathbf{w}_1' & \mathbf{w}_2' \\ \mathbf{w}_1 & \mathrm{diag}(\mathbf{w}_1) & \mathbf{I}_{12} \\ \mathbf{w}_2 & \mathbf{I}_{12}' & \mathrm{diag}(\mathbf{w}_2) \end{pmatrix}$$

with $\mathbf{w}_1 = (\delta_1(\{i\}))_{i=1,...I-1}$, $\mathbf{w}_2 = (\delta_2(\{j\}))_{j=1,...,J-1}$, and $\mathbf{I}_{12} = (\delta(\{(i,j)\}))_{i=1,...,I-1}^{j=1,...,J-1}$. By the strict convexity of the transformed criterion function $-\ln\det$ the information matrix of a D-optimum design is uniquely determined. Hence it is necessary for δ to be D-optimum that

$\delta(\{(i,j)\}) = \delta^*(\{(i,j)\}) = \frac{1}{IJ}$ for all treatment level combinations, $i = 1, ..., I-1$, $j = 1, ..., J-1$;

$\delta_1(\{i\}) = \delta_1^*(\{i\}) = \frac{1}{I}$, $i = 1, ..., I-1$, and

$\delta_2(\{j\}) = \delta_2^*(\{j\}) = \frac{1}{J}$, $j = 1, ..., J-1$,

for the marginal frequencies of the treatment levels.

These conditions imply that $\delta(\{(i,j)\}) = \frac{1}{IJ} = \delta^*(\{(i,j)\})$ for all level combinations (i,j), $i = 1, ..., I$, $j = 1, ..., J$. Hence the equireplicated design δ^* is the unique D-optimum design. □

Theorem 5.2 can immediately be extended to K factors as follows,

Corollary 5.3

Let δ_k^* be D-optimum in the single factor model $\mu_k(t_k) = \beta_0 + f_k(t_k)'\beta_k$, $t_k \in T_k$. Then the K-fold product design $\delta^* = \bigotimes_{k=1}^K \delta_k^*$ is D-optimum in the additive K-factor model $\mu(t_1, ..., t_K) = \beta_0 + \sum_{k=1}^K f_k(t_k)'\beta_k$, $(t_1, ..., t_K) \in T_1 \times ... \times T_K$.

Proof

For $K = 2$ the result has been shown in Theorem 5.2. For $K > 2$ we can identify the second up to the Kth factor as a new *meta*-factor $\widetilde{t}_2 = (t_2, ..., t_K) \in \widetilde{T}_2 = T_2 \times ... \times T_K$ with its marginal model

$$\widetilde{\mu}_2(\widetilde{t}_2) = \beta_0 + \widetilde{f}_2(\widetilde{t}_2)'\widetilde{\beta}_2$$

where $\widetilde{f}_2(\widetilde{t}_2) = (f_2(t_2)', ..., f_K(t_K)')'$. Thus we can consider the two-factor model

$$\mu(t_1, ..., t_K) = \widetilde{\mu}(t_1, \widetilde{t}_2) = \beta_0 + f_1(t_1)'\beta_1 + \widetilde{f}_2((t_2, ..., t_K))'\widetilde{\beta}_2 ,$$

$(t_1, \widetilde{t}_2) \in T_1 \times \widetilde{T}_2$. By Theorem 5.2 the product design $\delta_1^* \otimes \widetilde{\delta}_2^*$ is D-optimum if the marginal design $\widetilde{\delta}_2^*$ on $\widetilde{T}_2$ is D-optimum in the derived marginal model described by $\widetilde{\mu}_2$. Iterating back yields the desired result. □

As the number of supporting points for a product design is at least $p_1 p_2$ and hence considerably larger than the number $p_1 + p_2 - 1$ of parameters it is desirable to find optimum designs which have a smaller or even minimum number. That this is possible in some cases is exhibited in the following example,

Example 5.5 *K-way layout without interactions:*

$$\mu(i_1, i_2, ..., i_K) = \mu_0 + \sum_{k=1}^K \alpha_{i_k}^{(k)} ,$$

$i_k = 1, ..., I_k$, $k = 1, ..., K$. As in the special case of a two-way layout (see Example 5.1) we introduce the identifiability conditions $\alpha_{I_k}^{(k)} = 0$, $k = 1, ..., K$, of complete control subject to the control level combination $(I_1, ..., I_K)$.

Corollary 5.3 ensures that the design $\delta^* = \bigotimes_{k=1}^K \delta_k^*$ which assigns equal weights $\frac{1}{I_1 I_2 ... I_K}$ to each level combination $(i_1, ..., i_K)$, $i_k = 1, ..., I_k$, $k = 1, ..., K$, is D-optimum. For

$K > 2$ the D-optimum design is not necesseraly unique, because only the one- and two-dimensional marginals of the design are determined by the information matrix (see Example 5.4). Designs which may result in the same information matrix as a product design are known as Latin squares, orthogonal arrays, fractional factorials, or, more generally, as proportional designs (see RAKTOE, HEDAYAT, and FEDERER (1981), RAGHAVARAO (1971), BANDEMER et al. (1977), pp 388, and KUROTSCHKA (1978)).

To illustrate this feature we consider the three-way layout with each factor acting on two levels. The D-optimum product design δ^* which assigns equal weights $\frac{1}{8}$ to each of the 8 level combinations results in the information matrix

$$\mathbf{I}(\delta^*) = \frac{1}{4}\begin{pmatrix} 4 & 2 & 2 & 2 \\ 2 & 2 & 1 & 1 \\ 2 & 1 & 2 & 1 \\ 2 & 1 & 1 & 2 \end{pmatrix}.$$

The same information matrix can be obtained by a half-fraction $\widetilde{\delta}^*$ of δ^* which assigns equal weights $\frac{1}{4}$ to the four design points $(1,1,1)$, $(1,2,2)$, $(2,1,2)$ and $(2,2,1)$ and, hence, $\widetilde{\delta}^*$ is also D-optimum. □

This example suggests a complete characterization of D-optimum designs in K-factor models without interactions which generalizes the results obtained by WIERICH (1986c) for the K-way layout,

Corollary 5.4

δ^* is D-optimum in the additive K-factor model $\mu(t_1,...,t_K) = \beta_0 + \sum_{k=1}^K f_k(t_k)'\beta_k$, $(t_1,...,t_K) \in T_1 \times ... \times T_K$ if and only if

(i) the one-dimensional marginals δ_k^* of δ^* are D-optimum in the kth marginal models $\mu_k(t_k) = \beta_0 + f_k(t_k)'\beta_k$, $t_k \in T_k$, and

(ii) each pair of regression functions f_{k_1}, f_{k_2} is uncorrelated *with respect to the two-dimensional marginals δ_{k_1,k_2}^* of δ^* for the corresponding factors k_1, k_2, i. e. $\int f_{k_1} f_{k_2}'\, d\delta_{k_1 k_2}^* = \int f_{k_1}\, d\delta_{k_1}^* \int f_{k_2}'\, d\delta_{k_2}^*$, $1 \le k_1 < k_2 \le K$.*

Proof

The information matrix of a D-optimum design is uniquely determined. Hence by Corollary 5.3 we have

$$\mathbf{I}(\delta^*) = \mathbf{I}(\bigotimes_{k=1}^K \widetilde{\delta}_k)$$

for every D-optimum δ^*, where the $\widetilde{\delta}_k$ are D-optimum in the marginal models. In particular, this means $\int f_k\, d\delta_k^* = \int f_k\, d\delta^* = \int f_k\, d\widetilde{\delta}_k$, $\int f_k f_k'\, d\delta_k^* = \int f_k f_k'\, d\delta^* = \int f_k f_k'\, d\widetilde{\delta}_k$, such that $\mathbf{I}_k(\delta_k^*) = \mathbf{I}_k(\widetilde{\delta}_k)$, and

$$\int f_{k_1} f_{k_2}'\, d\delta_{k_1 k_2}^* = \int f_{k_1} f_{k_2}'\, d\delta^* = \int f_{k_1} f_{k_2}'\, d(\widetilde{\delta}_{k_1} \otimes \widetilde{\delta}_{k_2}) = \int f_{k_1}\, d\widetilde{\delta}_{k_1} \int f_{k_2}'\, d\widetilde{\delta}_{k_2},$$

$k, k_1, k_2 = 1, ..., K$, $k_1 \ne k_2$, which proves the direct part. Conversely (ii) implies $\mathbf{I}(\delta^*) = \mathbf{I}(\bigotimes_{k=1}^K \delta_k^*)$ and the D-optimality of δ^* follows from (i) by Corollary 5.3. □

In Corollary 5.4 the orthogonality condition (ii) is satisfied, for example, if the design δ under consideration is *second order proportional*, i. e. if each two-dimensional marginal design δ_{k_1,k_2} of δ is the product of the corresponding one-dimensional marginal designs $\delta_{k_1,k_2} = \delta_{k_1} \otimes \delta_{k_2}, 1 \le k_1 < k_2 \le K$ (for this definition see e.g. KUROTSCHKA (1978) or WIERICH (1989b), cf also RAGHAVARAO (1971)). Note that the design $\widetilde{\delta}^*$ considered in Example 5.5 is second order proportional.

Example 5.6 *K-way layout without interactions (complete control; cf Example 5.5):* Because of the special structure of the regression functions $f_k(i_k) = \mathbf{e}_{I_k-1,i_k}$ we can recover the one- and two-dimensional marginals δ_k and δ_{k_1,k_2} of every design δ from the entries in the information matrix (cf Example 5.4 for $K = 2$):

$$\int f_k d\delta = (\delta_k(\{i_k\}))_{i_k=1,...,I_k-1} \, ,$$
$$\int f_{k_1} f'_{k_2} d\delta = (\delta_{k_1,k_2}(\{(i_{k_1}, i_{k_2})\}))_{i_{k_1}=1,...,I_{k_1}-1}^{i_{k_2}=1,...,I_{k_2}-1} \, ,$$

$k, k_1, k_2 = 1, ..., K$, $k_1 \neq k_2$. Hence, condition (ii) of Corollary 5.4 is equivalent to $\delta_{k_1,k_2}(\{(i_{k_1}, i_{k_2})\}) = \delta_{k_1}(\{i_{k_1}\})\delta_{k_2}(\{i_{k_2}\})$, for all level combinations (i_{k_1}, i_{k_2}), which, in turn, is the second order proportionality of δ. This characterization of D-optimum designs for the K-way layout without interactions is due to WIERICH (1986c) (see also WIERICH (1986a)). □

For the other optimality criteria we need the following general representation of the covariance matrix $\mathbf{C}(\delta_1 \otimes \delta_2)$ of a product design $\delta_1 \otimes \delta_2 \in \Delta(\beta)$ in an additive two-factor model:

Lemma 5.5
(i) $\delta_1 \otimes \delta_2 \in \Delta(\beta)$ *if and only if* $\delta_k \in \Delta_k(\beta^{(k)})$, $k = 1, 2$.
(ii) *For every product design* $\delta_1 \otimes \delta_2 \in \Delta(\beta)$

$$\mathbf{C}(\delta_1 \otimes \delta_2) = \begin{pmatrix} \mathbf{C}_{\beta_0}(\delta_1 \otimes \delta_2) & -(\int f_1\, d\delta_1)'\mathbf{C}_{1,\beta_1}(\delta_1) & -(\int f_2\, d\delta_2)'\mathbf{C}_{2,\beta_2}(\delta_2) \\ -\mathbf{C}_{1,\beta_1}(\delta_1) \int f_1\, d\delta_1 & \mathbf{C}_{1,\beta_1}(\delta_1) & \mathbf{0} \\ -\mathbf{C}_{2,\beta_2}(\delta_2) \int f_2\, d\delta_2 & \mathbf{0} & \mathbf{C}_{2,\beta_2}(\delta_2) \end{pmatrix}$$

is the covariance matrix in the additive model and

$$\mathbf{C}_k(\delta_k) = \begin{pmatrix} \mathbf{C}_{k,\beta_0}(\delta_k) & -(\int f_k\, d\delta_k)'\mathbf{C}_{k,\beta_k}(\delta_k) \\ -\mathbf{C}_{k,\beta_k}(\delta_k) \int f_k\, d\delta_k & \mathbf{C}_{k,\beta_k}(\delta_k) \end{pmatrix}$$

is the covariance matrix of the marginal design δ_k *in the* k*th marginal model.*
The variances for β_0 *are related by* $\mathbf{C}_{\beta_0}(\delta_1 \otimes \delta_2) = \mathbf{C}_{1,\beta_0}(\delta_1) + \mathbf{C}_{2,\beta_0}(\delta_2) - 1$.

Proof
(i) This follows directly from the factorization $\det(\mathbf{I}(\delta_1 \otimes \delta_2)) = \det(\mathbf{I}_1(\delta_1)) \det(\mathbf{I}_2(\delta_2))$ (Lemma 5.1) which implies that $\mathbf{I}(\delta_1 \otimes \delta_2)$ is regular if and only if $\mathbf{I}_1(\delta_1)$ and $\mathbf{I}_2(\delta_2)$ are both regular.
(ii) Using the technique of orthogonalization (centering) with respect to the design introduced in Subsection 3.1 we obtain for the covariance matrix

$$\mathbf{C}(\delta_1 \otimes \delta_2) = \mathbf{I}(\delta_1 \otimes \delta_2)^{-1} = \widetilde{L}'\widetilde{\mathbf{I}}(\delta_1 \otimes \delta_2)^{-1}\widetilde{L}$$

of a product design $\delta_1 \otimes \delta_2 \in \Delta(\beta)$, where

$$\widetilde{L} = \begin{pmatrix} 1 & \mathbf{0} & \mathbf{0} \\ -\int f_1\,d\delta_1 & \mathbf{E}_{p_1-1} & \mathbf{0} \\ -\int f_2\,d\delta_2 & \mathbf{0} & \mathbf{E}_{p_2-1} \end{pmatrix}$$

is the transformation matrix for the orthogonolization and

$$\widetilde{\mathbf{I}}(\delta_1 \otimes \delta_2) = \begin{pmatrix} 1 & \mathbf{0} & \mathbf{0} \\ \mathbf{0} & \int f_1 f_1'\,d\delta_1 - \int f_1\,d\delta_1(\int f_1\,d\delta_1)' & \mathbf{0} \\ \mathbf{0} & \mathbf{0} & \int f_2 f_2'\,d\delta_2 - \int f_2\,d\delta_2(\int f_2\,d\delta_2)' \end{pmatrix}$$

is the information matrix for the transformed model, $\widetilde{a} = \widetilde{L}a$, $\widetilde{\beta} = \widetilde{L}'^{-1}\beta$. Analogously, we obtain for the marginal models

$$\mathbf{C}_k(\delta_k) = \mathbf{I}_k(\delta_k)^{-1} = \widetilde{L}_k'\widetilde{\mathbf{I}}_k(\delta_k)^{-1}\widetilde{L}_k$$

with

$$\widetilde{L}_k = \begin{pmatrix} 1 & \mathbf{0} \\ -\int f_k\,d\delta_k & \mathbf{E}_{p_k-1} \end{pmatrix}$$

and

$$\widetilde{\mathbf{I}}_k(\delta_k) = \begin{pmatrix} 1 & \mathbf{0} \\ \mathbf{0} & \int f_k f_k'\,d\delta_k - \int f_k\,d\delta_k(\int f_k\,d\delta_k)' \end{pmatrix}.$$

The block-diagonal matrices $\widetilde{\mathbf{I}}(\delta_1 \otimes \delta_2)$ and $\widetilde{\mathbf{I}}_k(\delta_k)$ can be inverted by inverting the blocks separately. Post- and pre-multiplication by $\widetilde{L}$, $\widetilde{L}_k$ and their transposes, respectively, yield the desired result.

Finally, the relationship between the variances for β_0 in the different models follows by Lemma 3.2. □

We notice from Lemma 5.5 that for a product design $\delta_1 \otimes \delta_2 \in \Delta(\beta)$ the covariance matrices for β_1 and β_2 in the additive model coincide with those in the corresponding marginal models, $\mathbf{C}_{\beta_k}(\delta_1 \otimes \delta_2) = \mathbf{C}_{k,\beta_k}(\delta_k)$. Moreover, for the linear aspect $\psi(\beta) = \binom{\beta_1}{\beta_2}$ we obtain a block diagonal covariance matrix

$$\mathbf{C}_\psi(\delta_1 \otimes \delta_2) = \begin{pmatrix} \mathbf{C}_{1,\beta_1}(\delta_1) & \mathbf{0} \\ \mathbf{0} & \mathbf{C}_{2,\beta_2}(\delta_2) \end{pmatrix},$$

such that the covariance matrix $\mathbf{C}_\psi(\delta_1 \otimes \delta_2)$ splits up into its marginal counterparts $\mathbf{C}_{k,\beta_k}(\delta_k)$ and the estimators for β_1 and β_2 are uncorrelated. This holds, more generally, for all linear aspects ψ with components depending only on the parameters associated with single factors, $\psi(\beta) = \binom{\psi_1(\beta_1)}{\psi_2(\beta_2)}$ (see Lemma 5.8).

We are now able to show that the best product design with respect to the A-criterion has marginals which are A-optimum in the marginal models:

Theorem 5.6

Let δ_k^* be A-optimum in the single factor model $\mu_k(t_k) = \beta_0 + f_k(t_k)'\beta_k$, $t_k \in T_k$. Then $\delta^* = \delta_1^* \otimes \delta_2^*$ is A-optimum within the class of product designs in the additive two-factor model $\mu(t_1, t_2) = \beta_0 + f_1(t_1)'\beta_1 + f_2(t_2)'\beta_2$, $(t_1, t_2) \in T_1 \times T_2$.

Proof

By Lemma 5.5 the trace of the covariance matrix can be represented by the traces of the corresponding covariance matrices in the marginal models according to

$$\begin{aligned}\operatorname{tr}(\mathbf{C}(\delta_1 \otimes \delta_2)) &= \mathbf{C}_{\beta_0}(\delta_1 \otimes \delta_2) + \operatorname{tr}(\mathbf{C}_{1,\beta_1}(\delta_1)) + \operatorname{tr}(\mathbf{C}_{2,\beta_2}(\delta_2)) \\ &= \operatorname{tr}(\mathbf{C}_1(\delta_1)) + \operatorname{tr}(\mathbf{C}_2(\delta_2)) - 1 \ .\end{aligned}$$

for every $\delta_1 \otimes \delta_2 \in \Delta(\beta)$. Hence, $\operatorname{tr}(\mathbf{C}(\delta_1 \otimes \delta_2))$ is minimized if $\operatorname{tr}(\mathbf{C}_1(\delta_1))$ and $\operatorname{tr}(\mathbf{C}_2(\delta_2))$ are minimized individually. □

A similar result holds for the Q-optimality if the weighting measure ξ is the product of the marginal weighting measures, $\xi = \xi_1 \otimes \xi_2$: For product designs $\delta_1 \otimes \delta_2 \in \Delta(\beta)$ the criterion function for the Q-criterion splits according to $\int a(t)'\mathbf{C}(\delta_1 \otimes \delta_2)a(t)\,\xi(dt) = \xi_2(T_2)\int a^{(1)}(t_1)'\mathbf{C}_1(\delta_1)a^{(1)}(t_1)\,\xi_1(dt_1)+\xi_1(T_1)\int a^{(2)}(t_2)'\mathbf{C}_2(\delta_2)a^{(2)}(t_2)\,\xi_2(dt_2)-\xi(T)$. Thus, $\delta_1^* \otimes \delta_2^*$ is Q-optimum with respect to $\xi = \xi_1 \otimes \xi_2$ in the additive two-factor model within the class of product designs if the designs δ_k^* are Q-optimum with respect to the marginal weighting measures ξ_k in the marginal models.

In the next example we see that there is no complete class result for product designs with respect to the A- and Q-criterion. Thus, in general, we cannot restrict our attention to product designs if we look for other criteria than the determinant although they will be highly efficient in many situations.

Example 5.7 *Two-way layout without interactions (complete control; cf Example 5.1):*

(i) By Theorem 5.6 the best product design with respect to the A-criterion is the product of the A-optimum designs for the marginal models (see WIERICH (1989b), Theorem 3 for this particular model). In the first marginal model the A-optimum design δ_1^* assigns equal weights $w_1^* = \frac{1}{I-1+\sqrt{I}}$ to the treatment levels $i = 1, ..., I-1$, and weight $1-(I-1)w_1^* = \frac{\sqrt{I}}{I-1+\sqrt{I}}$ to the control level resulting in a minimum value $\operatorname{tr}(\mathbf{C}_1(\delta_1^*)) = (I-1+\sqrt{I})^2$ for the trace of the marginal covariance matrix (cf Example 3.4). Analogously, for the second factor $\delta_2^*(\{j\}) = w_2^* = \frac{1}{J-1+\sqrt{J}}$, $j = 1, ..., J-1$, and $\delta_2^*(\{J\}) = \frac{\sqrt{J}}{J-1+\sqrt{J}}$ are the A-optimum weights and $\operatorname{tr}(\mathbf{C}_2(\delta_2^*)) = (J-1+\sqrt{J})^2$. As for the D-criterion $\delta_1^* \otimes \delta_2^*$ is the unique A-optimum design within the class of product designs. However, $\delta_1^* \otimes \delta_2^*$ is not A-optimum in Δ which can be seen by means of FEDOROV's equivalence theorem (Theorem 2.2) as follows (cf WIERICH (1989b)).

For every product design with marginals invariant with respect to permutations of the treatment levels we get

$$\mathbf{C}(\delta_1 \otimes \delta_2) = \begin{pmatrix} 1 + \frac{(I-1)w_1}{1-(I-1)w_1} + \frac{(J-1)w_2}{1-(J-1)w_2} & -\frac{1}{1-(I-1)w_1}\mathbf{1}_{I-1}' & -\frac{1}{1-(J-1)w_2}\mathbf{1}_{J-1}' \\ -\frac{1}{1-(I-1)w_1}\mathbf{1}_{I-1} & \frac{1}{w_1}\mathbf{E}_{I-1} + \frac{1}{1-(I-1)w_1}\mathbf{1}_{I-1}^{I-1} & \mathbf{0} \\ -\frac{1}{1-(J-1)w_2}\mathbf{1}_{J-1} & \mathbf{0} & \frac{1}{w_2}\mathbf{E}_{J-1} + \frac{1}{1-(J-1)w_2}\mathbf{1}_{J-1}^{J-1} \end{pmatrix}$$

by Lemma 5.5 (see Example 3.4 for $\mathbf{C}_k(\delta_k)$) where w_k is the marginal weight assigned to a treatment level in the kth marginal model. Hence, for a treatment level combination (i,j), $i = 1, ..., I-1$, $j = 1, ..., J-1$, we obtain $a(i,j)'\mathbf{C}(\delta_1 \otimes \delta_2) = (-1, \frac{1}{w_1}\mathbf{e}_{I-1,i}', \frac{1}{w_2}\mathbf{e}_{J-1,j}')$

and, consequently, $a(i,j)'\mathbf{C}(\delta_1 \otimes \delta_2)^2 a(i,j) = 1 + \frac{1}{w_1^2} + \frac{1}{w_2^2}$. For the optimum weights w_1^* and w_2^* this results in

$$a(i,j)'\mathbf{C}(\delta_1^* \otimes \delta_2^*)^2 a(i,j) = (I - 1 + \sqrt{I})^2 + (J - 1 + \sqrt{J})^2 + 1 = \mathrm{tr}(\mathbf{C}(\delta_1^* \otimes \delta_2^*)) + 2$$

where the last equality follows from Lemma 5.5. Thus, $\delta_1^* \otimes \delta_2^*$ does not fulfil the conditions of FEDOROV's equivalence theorem which shows that the product design $\delta_1^* \otimes \delta_2^*$ is not A-optimum.

(ii) For the Q-criterion we consider the situation of two levels for each factor ($I = J = 2$):

$$\mu(i,j) = \begin{cases} \mu_0 + \alpha^{(1)} + \alpha^{(2)} & , i = j = 1, \\ \mu_0 + \alpha^{(1)} & , i = 1, j = 2, \\ \mu_0 + \alpha^{(2)} & , i = 2, j = 1, \\ \mu_0 & , i = j = 2. \end{cases}$$

Let the weighting measures $\xi_1 = \xi_2$ be given by $\xi_1(\{1\}) = v = 1 - \xi_1(\{2\})$. Then the Q-optimum marginal designs δ_k^* with respect to ξ_k are determined by $\delta_k^*(\{1\}) = w^*$ and $\delta_k^*(\{2\}) = 1 - w^*$ where the optimum weight w^* equals $w^* = \frac{\sqrt{v}}{\sqrt{v} + \sqrt{1-v}}$. We note that $w^* \neq v$ if $v \notin \{0, \frac{1}{2}, 1\}$.

For identical marginal designs $\delta_1 = \delta_2$ the covariance matrix of the product design is

$$\mathbf{C}(\delta_1 \otimes \delta_2) = \frac{1}{w(1-w)} \begin{pmatrix} w^2 + w & -w & -w \\ -w & 1 & 0 \\ -w & 0 & 1 \end{pmatrix}$$

in this particular model with two-levels for each factor, $w = \delta_k(\{1\})$, and the marginal covariance matrices are

$$\mathbf{C}_k(\delta_k) = \frac{1}{w(1-w)} \begin{pmatrix} w & -w \\ -w & 1 \end{pmatrix}.$$

Let us assume that $0 < v < 1$ (the singular cases $v = 1$ and $v = 0$ can be treated by Theorem 5.15 or by Corollary 5.16), hence $0 < w^* < 1$ and $\delta_1^* \in \Delta_1(\beta^{(1)})$. For the treatment level combinations we obtain

$$\begin{aligned} a(1,1)'\mathbf{C}(\delta_1^* \otimes \delta_2^*) \textstyle\int aa'd\xi\ \mathbf{C}(\delta_1^* \otimes \delta_2^*) a(1,1) &= 3 - 2v + 4v\sqrt{v(1-v)} \\ \geq 1 + 4\sqrt{v(1-v)} &= \mathrm{tr}(\textstyle\int aa'\, d\xi\ \mathbf{C}(\delta_1^* \otimes \delta_2^*)) \end{aligned}$$

with equality if and only if $v = \frac{1}{2}$. Hence, by the equivalence theorem for Q-optimality (Corollary 2.8), the best product design $\delta_1^* \otimes \delta_2^*$ is not Q-optimum within the class Δ of all designs if $v \neq \frac{1}{2}$. (For $v = \frac{1}{2}$ see the results obtained in Subsection 5.2.) □

In case of E-optimality even less is known, because the product of E-optimum designs in the marginal models is not necessarily the best product design with respect to the E-criterion,

Example 5.8 *Two-way layout without interactions (complete control; cf Example 5.1):*
As in Example 5.7 (ii) we consider the situation of two levels for each factor. We can restrict to identical marginal designs $\delta_1 = \delta_2$ by consideration of invariance with respect

to permutations of the factors (see the remark preceding Example 3.5): $\delta_k(\{1\}) = w$, $\delta_k(\{2\}) = 1 - w$. The eigenvalues of the covariance matrix $\mathbf{C}(\delta_1 \otimes \delta_2)$ are $\frac{1}{2w(1-w)}(w^2 + w + 1 \pm \sqrt{(w^2+w+1)^2 - 4w(1-w)})$ and $\frac{1}{w(1-w)}$, and the maximum eigenvalue attains its minimum $\lambda_{\max}(w^*) = 5.70$ for the weight $w^* = \frac{1}{4}(\sqrt{41} - 5) \approx 0.35$. This optimum weight w^* differs from the corresponding E-optimum choice $w_1^* = \frac{2}{5} = 0.40$ (see Example 3.4) which results in a product design with maximum eigenvalue $\lambda_{\max}(w_1^*) = 5.78$. □

In many practical situations it is not of interest to make inference on the intercept term because this constant may represent some general environmental conditions which are kept fixed throughout the whole experiment, but which may vary substantially for forthcoming observations. Therefore we draw our attention to parts of the whole parameter vector excluding the general mean.

For parameters associated with the effects of a single factor, e. g. β_1, the optimum designs can be determined in a straightforward way by using refinement arguments of Subsection 3.1 because the corresponding marginal model can be regarded as a submodel of the whole additive model,

Lemma 5.7

Let ψ be a linear aspect of β, $\psi(\beta) = L_\psi \beta \in \mathbb{R}^s$, let $L_\psi = (L_0|L_1|L_2)$ be partitioned according to the influence of the constant term and the single factors on ψ, i. e. $L_0 \in \mathbb{R}^s$, $L_k \in \mathbb{R}^{s\times(p_k-1)}$, and let ψ_k be the associated linear aspect in the marginal model with $L_{k,\psi_k} = (L_0|L_k)$.

(i) If $\delta_1 \in \Delta_1(\psi_1)$ and $\delta_2 \in \Delta_2(\psi_2)$ then $\delta_1 \otimes \delta_2 \in \Delta(\psi)$.

(ii) If $\delta \in \Delta(\psi)$ then $\delta_k \in \Delta_k(\psi_k)$ for the marginals of δ and, hence, $\delta_1 \otimes \delta_2 \in \Delta(\psi)$.

(iii) Let ψ be a linear aspect depending on β_1, i. e. $\psi(\beta) = \psi_1(\beta^{(1)}) = L_1\beta_1$. Then $\delta_1 \in \Delta_1(\psi_1)$ implies $\delta_1 \otimes \delta_2 \in \Delta(\psi)$ for every marginal design δ_2 on T_2.

Proof

For $\delta_k \in \Delta_k(\psi_k)$ there is a matrix $M^{(k)}$ with $L_{k,\psi_k} = M^{(k)}\mathbf{I}_k(\delta_k)$ which can be partitioned in the same way as L_{k,ψ_k}, $M^{(k)} = (M_0^{(k)}|M_1^{(k)})$, $M_0^{(k)} \in \mathbb{R}^s$, $M_1^{(k)} \in \mathbb{R}^{s\times(p_k-1)}$.

(i) Let $M = (M_0^{(1)} + M_0^{(2)} - L_0|M_1^{(1)}|M_1^{(2)})$, then we obtain $M\mathbf{I}(\delta_1 \otimes \delta_2) = L_\psi$ and, hence, $\delta_1 \otimes \delta_2 \in \Delta(\psi)$.

(ii) If we regard the kth marginal model as a submodel of the additive model the result follows immediately from the refinement argument of Lemma 3.5 and (i).

(iii) Consider the degenerate linear aspect $\psi_2 = \mathbf{0}$ and let $M = (M^{(1)}|\mathbf{0})$, then $L_\psi = M\mathbf{I}(\delta_1 \otimes \delta_2)$ which proves the identifiability. □

In case of an underlying orthogonality the next lemma presents a representation for the generalized inverse of the information matrix $\mathbf{I}(\delta)$ in the additive model which substantially simplifies the calculation of covariance matrices for identifiable aspects and which extends the regular case for product designs treated in Lemma 5.5.

Lemma 5.8

If f_1 and f_2 are uncorrelated *with respect to δ, i. e. $\int f_1 f_2' \, d\delta = \int f_1' \, d\delta_1 \int f_2' \, d\delta_2$, then*

$$\mathbf{I}(\delta)^- = \begin{pmatrix} 1 + \int f_1' \, d\delta_1 J_1^- \int f_1 \, d\delta_1 + \int f_2' \, d\delta_2 J_2^- \int f_2 \, d\delta_2 & -\int f_1' \, d\delta_1 J_1^- & -\int f_2' \, d\delta_2 J_2^- \\ -J_1^- \int f_1 \, d\delta_1 & J_1^- & \mathbf{0} \\ -J_2^- \int f_2 \, d\delta_2 & \mathbf{0} & J_2^- \end{pmatrix}$$

is a generalized inverse of the information matrix $\mathbf{I}(\delta)$, *where* δ_k *is the kth marginal of* δ *and* $J_k = \int f_k f_k' \, d\delta_k - \int f_k \, d\delta_k \int f_k' \, d\delta_k$.

Proof

Following the proof of Lemma 5.5 the result is a straightforward generalization to possibly singular information matrices. By the orthogonalization technique of Subsection 3.1 we obtain $\mathbf{I}(\delta) = \widetilde{L}^{-1}\widetilde{\mathbf{I}}(\delta)\widetilde{L}'^{-1}$ with a block diagonal transformed information matrix

$$\widetilde{\mathbf{I}}(\delta) = \begin{pmatrix} 1 & \mathbf{0} & \mathbf{0} \\ \mathbf{0} & J_1 & \mathbf{0} \\ \mathbf{0} & \mathbf{0} & J_2 \end{pmatrix}$$

and the same regular transformation matrix $\widetilde{L}$ as in the proof of Lemma 5.5. As

$$J = \begin{pmatrix} J_1 & \mathbf{0} \\ \mathbf{0} & J_2 \end{pmatrix}$$

can be inverted blockwise the representation of $\mathbf{I}(\delta)^-$ follows from Lemma 3.2. □

The condition of uncorrelated regresion functions is always satisfied for product designs. With these preparatory results we are able to present a complete class theorem for all situations in which we are interested in the parameters associated with a single factor and the other factor may be considered as a noise factor including e. g. block effects,

Theorem 5.9

Let ψ *be a linear aspect depending on the partial parameter vector* β_1, $\psi(\beta) = L_1\beta_1$, *in the additive model* $\mu(t_1, t_2) = \beta_0 + f_1(t_1)'\beta_1 + f_2(t_2)'\beta_2$, $(t_1, t_2) \in T_1 \times T_2$. *Then for the covariance matrices* $\mathbf{C}_\psi(\delta) \geq \mathbf{C}_\psi(\delta_1 \otimes \delta_2)$ *for every design* $\delta \in \Delta(\psi)$, *where* δ_1 *and* δ_2 *are the marginals of* δ.

In particular, $\mathbf{C}_{\beta_1}(\delta) \geq \mathbf{C}_{\beta_1}(\delta_1 \otimes \delta_2)$ *for* $\delta \in \Delta(\beta_1)$, *with equality if and only if* f_1 *and* f_2 *are uncorrelated with respect to* δ.

Proof

As the first marginal model can be considered as a submodel the refinement argument of Lemma 3.5 shows $\mathbf{C}_\psi(\delta) \geq \mathbf{C}_{1,\psi_1}(\delta_1)$, where $\psi_1(\beta^{(1)}) = L_1\beta_1$ is the associated linear aspect in the marginal model. The selection matrices for ψ and ψ_1 are $L_\psi = (\mathbf{0}|L_1|\mathbf{0})$ and $L_{1,\psi_1} = (\mathbf{0}|L_1)$ and, hence, $\mathbf{C}_\psi(\delta_1 \otimes \delta_2) = \mathbf{C}_{1,\psi_1}(\delta_1)$ in view of Lemma 5.8. □

For marginal designs δ_1 and δ_2 (supported on a finite set) the resulting estimators $\widehat{\psi}_{\delta_1\otimes\delta_2}$ and $\widehat{\psi}_{1,\delta_1}$ coincide for the same linear aspects ψ and ψ_1 depending on β_1, $\psi(\beta) = \psi_1(\beta^{(1)}) = L_1\beta_1$, in both the additive model and the associated marginal model. Also the covariance matrices coincide and the performance of the inference on $L_1\beta_1$ does not depend on the choice of the correct model if a product design is used.

By interchanging factors it is obvious that the above results are also valid for the second factor in case ψ depends on β_2. Note also that a reparametrization of that part of the regression function which is associated with the other marginal model causes only a transformation of the remaining parameters, say, $\binom{\beta_0}{\beta_1}$ and does not affect the covariance matrices for β_2,

Corollary 5.10

Let ψ be a linear aspect depending on the partial parameter vector β_2, $\psi(\beta) = L_2\beta_2$ in the two-factor model

$$\mu(t_1, t_2) = a^{(1)}(t_1)'\beta_1 + f_2(t_2)'\beta_2 , \tag{5.4}$$

$(t_1, t_2) \in T_1 \times T_2$, without interactions. If $\mathbf{1} \in \mathrm{span}(a_1^{(1)}, ..., a_{p_1}^{(1)})$, i. e. the constant function is a linear combination of the regression functions in the first marginal model, then for the covariance matrices $\mathbf{C}_\psi(\delta) \geq \mathbf{C}_\psi(\delta_1 \otimes \delta_2)$ for every design $\delta \in \Delta(\psi)$, where δ_1 and δ_2 are the marginals of δ.

In particular, $\mathbf{C}_{\beta_2}(\delta) \geq \mathbf{C}_{\beta_2}(\delta_1 \otimes \delta_2)$ for $\delta \in \Delta(\beta_2)$, with equality if and only if $\int a^{(1)} f_2' \, d\delta = \int a^{(1)} \, d\delta_1 \int f_2' \, d\delta_2$.

Proof

If $\mathbf{1} \in \mathrm{span}(a_1^{(1)}, ..., a_{p_1}^{(1)})$ there is a one-to-one linear transformation such that $\widetilde{a}^{(1)} = \binom{1}{f_1} = L_1 a^{(1)}$. The new parametrization is that of an additive model considered so far, $\mu(t_1, t_2) = \widetilde{\beta}_0 + f_1(t_1)'\widetilde{\beta}_1 + f_2(t_2)'\beta_2$. Let

$$L = \begin{pmatrix} L_1 & \mathbf{0} \\ \mathbf{0} & \mathbf{E}_{p_1 - 1} \end{pmatrix}$$

be the corresponding transformation matrix in the two-factor model, then the associated transformation $\widetilde{\beta} = L'^{-1}\beta$ of the parameters leaves β_2 unchanged, $\psi(\widetilde{\beta}) = \psi(\beta)$, and the covariance matrices $\mathbf{C}_\psi$ coincide in both the transformed and the original model. Hence, the result follows directly from Theorem 5.9.

To obtain the final equivalence from that of Theorem 5.9 we note that $\int a^{(1)} f_2' \, d\delta = \int a^{(1)} \, d\delta_1 \int f_2' \, d\delta_2$ if and only if $\int f_1 f_2' \, d\delta = \int f_1 \, d\delta_1 \int f_2' \, d\delta_2$. □

With these results the search for optimum designs can be restricted to product designs for every optimality criterion based on β_1.

Corollary 5.11

Let ψ depend on β_1, $\psi(\beta) = \psi_1(\beta^{(1)}) = L_1\beta_1$. Then the product design $\delta^* = \delta_1^* \otimes \delta_2$ is D_ψ- (resp. A_ψ-) optimum in the additive model $\mu(t_1, t_2) = \beta_0 + f_1'\beta_1 + f_2'\beta_2$, $(t_1, t_2) \in T_1 \times T_2$, for every marginal design δ_2 on T_2, if and only if δ_1^* is D_{ψ_1}- (resp. A_{ψ_1}-) optimum in the first marginal model $\mu_1(t_1) = \beta_0 + f_1(t_1)'\beta_1$, $t_1 \in T_1$.

Proof

The result is a straightforward consequence of Theorem 5.9 because $\delta_1 \in \Delta_1(\psi_1)$ if and only if $\delta_1 \otimes \delta_2 \in \Delta(\psi)$ in view of Lemma 5.7. □

In particular, for $\psi(\beta) = \beta_1$ we obtain that a product design $\delta_1^* \otimes \delta_2$ is D_{β_1}- resp. A_{β_1}-optimum if and only if δ_1^* is D_{β_1}- resp. A_{β_1}-optimum in the first marginal model.

Let ψ depend on β_2, $\psi(\beta) = \psi_2(\beta^{(2)}) = L_2\beta_2$. Then, by Corollary 5.10, the analogous result holds that the product design $\delta^* = \delta_1 \otimes \delta_2^*$ D_ψ- (resp. A_ψ-) optimum in the model (5.4) for every marginal design δ_1 on T_1, if and only if δ_2^* is D_{ψ_2}- (resp. A_{ψ_2}-) optimum in the second marginal model $\mu_2(t_2) = \beta_0 + f_2(t_2)'\beta_2$.

Example 5.9 *K-way layout without interactions (complete control; cf Example 5.5):*
Let $\beta_1 = (\alpha_i^{(1)})_{i=1,...,I_1-1}$ be the treatment effects of the first factor compared to control. Then $\mathbf{C}_{\beta_1}(\delta) \geq \mathbf{C}_{\beta_1}(\delta_1 \otimes \delta_{2,...,K})$ by Theorem 5.9, where $\delta_{2,...,K}$ is the $(K-1)$-dimensional marginal design on the $K-1$ factors 2 to K. Let $f_{2,...,K}$ be defined by $f_{2,...,K}(t_2, ..., t_K) = (f_2(t_2)', ..., f_K(t_K)')'$. Then equality is attained for the covariance matrices if and only if $\int f_1 f'_{2,...,K}\, d\delta = \int f_1\, d\delta_1 \int f'_{2,...,K}\, d\delta_{2,...,K}$ which becomes $\delta_{1,k}(\{(i_1, i_k)\}) = \delta_1(\{i_1\})\delta_k(\{i_k\})$ for all treatment level combinations $i_1 = 1, ..., I_1 - 1$, $i_k = 1, ..., I_k - 1$, $k = 2, ..., K$. This implies $\delta_{1,k}(\{(i_1, I_k)\}) = \delta_1(\{i_1\}) - \sum_{i_k=1}^{I_k-1} \delta_{1,k}(\{(i_1, i_k)\}) = \delta_1(\{i_1\})\delta_k(\{I_k\})$ for all treatment levels $i_1 = 1, ..., I_1 - 1$, and all control levels I_k, $k = 2, ..., K$. Analogously, $\delta_{1,k}(\{(I_1, i_k)\}) = \delta_1(\{I_1\})\delta_k(\{i_k\})$ for those level combinations in which the control level I_1 of the first factor is used. Combining these conditions we see that $\mathbf{C}_{\beta_1}(\delta) = \mathbf{C}_{1,\beta_1}(\delta_1)$ if and only if every two-dimensional marginal involving the first factor is the product of the corresponding one-dimensional marginals $\delta_{1,k} = \delta_1 \otimes \delta_k$, $k = 2, ..., K$.

Because of the strict convexity of the (transformed) D- and A-criterion we can conclude that δ^* is D- (resp. A-) optimum for the effects $\beta_1 = (\alpha_i^{(1)})_{i=1,...,I_1-1}$ of the first factor compared to control in the K-way layout without interactions if and only if

(i) δ_1^* is D- (resp. A) -optimum for the effects $\beta_1 = (\alpha_i)_{i=1,...,I_1-1}$ compared to control in the one-way layout $\mu_1(i) = \mu + \alpha_i$, $i = 1, ..., I_1$, with control ($\alpha_I^{(1)} = 0$); and

(ii) $\delta_{1,k}^* = \delta_1^* \otimes \delta_k^*$, $k = 2, ..., K$.

We recall that the D-optimum design for the effects compared to control is given by the weights $\delta_1^*(\{i\}) = \frac{1}{I}$ for all levels $i = 1, ..., I$ and the A-optimum design by the weights $\delta_1^*(\{i\}) = \frac{1}{I-1+\sqrt{I-1}}$ for the treatment levels $i = 1, ..., I-1$ and the weight $\delta_1^*(\{I\}) = \frac{\sqrt{I-1}}{I-1+\sqrt{I-1}}$ for the control level (see Example 3.6). We, thus, recover Theorem 4 of KUROTSCHKA (1978). □

As a direct consequence of Theorem 5.9 we derive the optimality of product designs if the linear aspect ψ consists of parts $\psi^{(1)}$ and $\psi^{(2)}$ depending separately on β_1 and on β_2 and if the combined criterion for ψ is a monotonic function of the marginal criteria. We exhibit this in case of the A-criterion, next,

Corollary 5.12
Let $\psi = \binom{\psi^{(1)}}{\psi^{(2)}}$ be a linear aspect given by $\psi^{(k)}(\beta) = L_k\beta_k$ and let ψ_k be the associated linear aspects in the kth marginal model, $\psi_k(\beta^{(k)}) = L_k\beta_k$. If δ_k^* is A_{ψ_k}-optimum in the single factor model $\mu_k(t_k) = \beta_0 + f_k(t_k)'\beta_k$, $t_k \in T_k$, then the product design $\delta^* = \delta_1^* \otimes \delta_2^*$ is A_ψ-optimum in the additive model $\mu(t_1, t_2) = \beta_0 + f_1(t_1)'\beta_1 + f_2(t_2)'\beta_2$, $(t_1, t_2) \in T_1 \times T_2$.

Proof
By Corollary 5.11 the product design $\delta_1^* \otimes \delta_2^*$ is simultaneously $A_{\psi^{(1)}}$- and $A_{\psi^{(2)}}$-optimum. Then the A_ψ-optimality follows from $\mathrm{tr}(\mathbf{C}_\psi(\delta)) = \mathrm{tr}(\mathbf{C}_{\psi^{(1)}}(\delta)) + \mathrm{tr}(\mathbf{C}_{\psi^{(2)}}(\delta))$ for every $\delta \in \Delta(\psi)$. □

In particular, if δ_1^* and δ_2^* are A_{β_1}- and A_{β_2}-optimum, then the product design $\delta^* = \delta_1^* \otimes \delta_2^*$ is A-optimum for $\binom{\beta_1}{\beta_2}$.

Example 5.10 *K-way layout without interactions (complete control; cf Example 5.5):*
For any set of M factors $(1 < M \leq K)$ we are going to determine those designs which are A-optimum for the set of parameters associated with the effects of these M factors.

Without loss of generality we may assume that the first M factors are of interest and the linear aspect is given by $\psi(\beta) = (\beta_1', ..., \beta_M')'$. Then by Corollary 5.11 the product $\widetilde{\delta} = \widetilde{\delta}_{1,...,M} \otimes \delta_{M+1,...,K}$ is A_ψ-optimum if $\widetilde{\delta}_{1,...,M}$ is A-optimum for the associated linear aspect $\psi_{(M)}$, $\psi_{(M)}((\beta_0, \beta_1', ..., \beta_M')') = (\beta_1', ..., \beta_M')'$, in the M-way layout

$$\mu_{(M)}(i_1, ..., i_M) = \mu_0 + \sum_{k=1}^{M} \alpha_{i_k}^{(k)},$$

without interactions. From Corollary 5.12 we obtain by induction that the product design $\widetilde{\delta}_{1,...,M} = \bigotimes_{k=1}^{M} \widetilde{\delta}_k$ is A-optimum for $\psi_{(M)}$ in the submodel if each marginal design $\widetilde{\delta}_k$ is A_{β_k}-optimum in the kth marginal model. Hence, we can conclude that the product $\widetilde{\delta} = \bigotimes_{k=1}^{M} \widetilde{\delta}_k \otimes \delta_{M+1,...,K}$ is A_ψ-optimum for each design $\delta_{M+1,...,K}$ on $T_{M+1} \times ... \times T_K$.

We recall that, according to Example 3.6, the unique A_{β_k}-optimum design $\widetilde{\delta}_k$ is given by $\widetilde{\delta}_k(\{i_k\}) = \frac{1}{I_k-1+\sqrt{I_k-1}}$, for the treatment levels $i_k = 1, ..., I_k - 1$, and $\widetilde{\delta}_k(\{I_k\}) = \frac{\sqrt{I_k-1}}{I_k-1+\sqrt{I_k-1}}$ for the control levels, $k = 1, ..., K$. Because of the strict convexity of the A_ψ-criterion as a function on the set $\{\mathbf{C}_\psi(\delta)^{-1}; \delta \in \Delta(\psi)\}$ the covariance matrix $\mathbf{C}_\psi(\delta^*)$ is uniquely determined by $\mathbf{C}_\psi(\delta^*) = \mathbf{C}_\psi(\widetilde{\delta})$. for every A_ψ-optimum design δ^* Denote the m-dimensional marginals of δ^* by $\delta^*_{k_1,...,k_m}$. Then $\mathbf{C}_\psi(\delta^*) = \mathbf{C}_\psi(\widetilde{\delta})$ if and only if $\mathbf{C}_\psi(\delta^*) = \mathbf{C}_{\psi_{(M)}}(\delta^*_{1,...,M}) = \mathbf{C}_{\psi_{(M)}}(\bigotimes_{k=1}^{M} \widetilde{\delta}_k)$ and the first equality holds if and only if $\int f_{k_1} f_{k_2}' \, d\delta^*_{k_1,k_2} = \int f_{k_1} \, d\delta^*_{k_1} (\int f_{k_2} \, d\delta^*_{k_2})'$ for every $k_1 = 1, ..., M$, and $k_2 = M + 1, ..., K$, because of Theorem 5.9. By repeated application of Theorem 5.9 the second equality is seen to be equivalent to $\mathbf{C}_{k,\beta_k}(\delta^*_k) = \mathbf{C}_{k,\beta_k}(\widetilde{\delta}_k)$ and $\int f_{k_1} f_{k_2}' \, d\delta^*_{k_1,k_2} = \int f_{k_1} \, d\delta^*_{k_1} (\int f_{k_2} \, d\delta^*_{k_2})'$, $k, k_1, k_2 = 1, ..., M$, $k_1 \neq k_2$.

Now, by the particular structure $f_k(i_k) = \mathbf{e}_{I_k-1,i_k}$ of the regression functions f_k we have $\int f_{k_1} f_{k_2}' \, d\delta^*_{k_1,k_2} = \int f_{k_1} \, d\delta^*_{k_1} \int f_{k_2}' \, d\delta^*_{k_2}$ if and only if $\delta^*_{k_1,k_2} = \delta^*_{k_1} \otimes \delta^*_{k_2}$ (cf Example 5.6). Combining all these equivalences we obtain that δ^* is A_ψ-optimum for the effects $(\alpha_{i_k}^{(k)}; i_k = 1, ..., I_k - 1, k = 1, ..., M)$ of the first M factors compared to control in the K-way layout without interactions if and only if

(i) $\delta^*_k(\{i_k\}) = \frac{1}{I_k-1+\sqrt{I_k-1}}$, $i_k = 1, ..., I_k - 1$,

$\delta^*_k(\{I_k\}) = \frac{\sqrt{I_k-1}}{I_k-1+\sqrt{I_k-1}}$ for $k = 1, ..., M$; and

(ii) $\delta^*_{k_1,k_2} = \delta^*_{k_1} \otimes \delta^*_{k_2}$ for $k_1 = 1, ..., M$, $k_2 = k_1 + 1, ..., K$.

This reproves the main Theorem of WIERICH (1988b) in view of Corollary 5.10. □

In the proof of Corollary 5.12 a complete class result has been implicitly used for the particular linear aspects ψ, $\mathrm{tr}(\mathbf{C}_\psi(\delta)) \geq \mathrm{tr}(\mathbf{C}_\psi(\delta_1 \otimes \delta_2))$. This can be extended to other criteria based on the eigenvalues in ordet to establish the following main result of this section.

Theorem 5.13

Let $\psi = \binom{\psi^{(1)}}{\psi^{(2)}}$ be a linear aspect given by $\psi^{(k)}(\beta) = L_k\beta_k$ and let ψ_k be the associated linear aspects in the kth marginal model, $\psi_k(\beta^{(k)}) = L_k\beta_k$. If δ^*_k is Φ_q-optimum for ψ_k in the single factor model $\mu_k(t_k) = \beta_0 + f_k(t_k)'\beta_k$, $t_k \in T_k$, then the product design $\delta^* = \delta^*_1 \otimes \delta^*_2$ is Φ_q-optimum for ψ in the additive model $\mu(t_1, t_2) = \beta_0 + f_1(t_1)'\beta_1 + f_2(t_2)'\beta_2$, $(t_1, t_2) \in T_1 \times T_2$.

Proof

For $\delta \in \Delta$ denote the marginals by δ_1 and δ_2. In analogy to the information matrix

$$\mathbf{I}(\delta) = \begin{pmatrix} 1 & (\int f_1\, d\delta_1)' & (\int f_2\, d\delta_2)' \\ \int f_1\, d\delta_1 & \int f_1 f_1'\, d\delta_1 & \int f_1 f_2'\, d\delta \\ \int f_2\, d\delta_2 & \int f_2 f_1'\, d\delta & \int f_2 f_2'\, d\delta_2 \end{pmatrix}$$

we define formally the partially sign changed matrix

$$\bar{\mathbf{I}}(\delta) = \begin{pmatrix} 1 & \int f_1'\, d\delta_1 & -\int f_2'\, d\delta_2 \\ \int f_1\, d\delta_1 & \int f_1 f_1'\, d\delta_1 & -\int f_1 f_2'\, d\delta \\ -\int f_2\, d\delta_2 & -\int f_2 f_1'\, d\delta & \int f_2 f_2'\, d\delta_2 \end{pmatrix} .$$

If we write $f(t_1, t_2) = \binom{f_1(t_1)}{f_2(t_2)}$ for the non-constant regression functions, then this formal sign change is accomplished by a transformation $\bar{f} = \bar{L} f$ with an associated orthogonal transformation matrix

$$\bar{L} = \begin{pmatrix} \mathbf{E}_{p_1 - 1} & \mathbf{0} \\ \mathbf{0} & -\mathbf{E}_{p_2 - 1} \end{pmatrix} .$$

By Lemma 3.2 the matrix

$$\mathbf{I}(\delta)^- = \begin{pmatrix} 1 + \int f'\, d\delta J(\delta)^- \int f\, d\delta & -\int f'\, d\delta J(\delta)^- \\ -J(\delta)^- \int f\, d\delta & J(\delta)^- \end{pmatrix}$$

is a generalized inverse of the information matrix $\mathbf{I}(\delta)$ and the matrix

$$\bar{\mathbf{I}}(\delta)^- = \begin{pmatrix} 1 + \int f'\, d\delta J(\delta)^- \int f\, d\delta & -\int f'\, d\delta J(\delta)^- \bar{L} \\ -\bar{L} J(\delta)^- \int f\, d\delta & \bar{L} J(\delta)^- \bar{L} \end{pmatrix}$$

is a generalized inverse of $\bar{\mathbf{I}}(\delta)$, where $J(\delta) = \int f f'\, d\delta - \int f\, d\delta \int f'\, d\delta$. Moreover,

$$J(\delta)^- = \begin{pmatrix} G_1 & G_{12} \\ G_{12}' & G_2 \end{pmatrix}$$

is a generalized inverse of $J(\delta)$ if and only if

$$\bar{J}(\delta)^- = \begin{pmatrix} G_1 & -G_{12} \\ -G_{12}' & G_2 \end{pmatrix}$$

is a generalized inverse of $\bar{J}(\delta) = \bar{L} J(\delta) \bar{L}$, where the matrices are partitioned appropriately according to the dimensions of f_1 and f_2. In analogy to the covariance matrix

$$\mathbf{C}_\psi(\delta) = L_\psi \mathbf{I}(\delta)^- L_\psi' = \begin{pmatrix} L_1 G_1 L_1' & L_1 G_{12} L_2' \\ (L_1 G_{12} L_2')' & L_2 G_2 L_2' \end{pmatrix}$$

for ψ we compute formally the covariance matrix

$$\bar{\mathbf{C}}_\psi(\delta) = L_\psi \bar{\mathbf{I}}(\delta)^- L_\psi' = \begin{pmatrix} L_1 G_1 L_1' & -L_1 G_{12} L_2' \\ -(L_1 G_{12} L_2')' & L_2 G_2 L_2' \end{pmatrix}$$

for ψ in case of a partial sign change, for every $\delta \in \Delta(\psi)$. The eigenvalues of $\mathbf{C}_\psi(\delta)$ and $\bar{\mathbf{C}}_\psi(\delta)$ coincide because $\mathbf{C}_\psi(\delta) z = \lambda z$ if and only if $\bar{\mathbf{C}}_\psi(\delta) \bar{z} = \lambda \bar{z}$, where the eigenvectors

$z = \binom{z_1}{z_2}$ and $\bar{z} = \binom{z_1}{-z_2}$ are appropriately partitioned according to the two factors, $z_k \in \mathbb{R}^{s_k}$. Hence, the criterion function φ_q of the Φ_q-criterion for ψ evaluated at $\mathbf{C}_\psi(\delta)$ and formally at $\bar{\mathbf{C}}_\psi(\delta)$ yields the same value showing an invariance with respect to the partial sign change.

The criterion function $\Phi_{q;\psi}$ of the Φ_q-criterion for ψ, $\Phi_{q;\psi}(\delta) = \varphi_q(\mathbf{C}_\psi(\delta))$, can be written as $\Phi_{q;\psi}(\delta) = h_{q,\psi}(\phi_{q;\psi}(J(\delta)))$ where $\phi_{q;\psi}$ is a convex function on the set of positive semi-definite matrices, $\phi_{q;\psi}$ depends on $J(\delta)$ via the eigenvalues of $\mathbf{C}_\psi(\delta)$ and $h_{q,\psi}$ is a monotonically increasing function (see e.g. KIEFER (1974)). Because of $\phi_{q;\psi}(J(\delta)) = \phi_{q;\psi}(\bar{J}(\delta))$ we obtain $\phi_{q;\psi}(J(\delta)) \geq \phi_{q;\psi}(\frac{1}{2}(J(\delta) + \bar{J}(\delta)))$ by the convexity of $\phi_{q;\psi}$. The identity $\frac{1}{2}(J(\delta)+\bar{J}(\delta)) = J(\delta_1 \otimes \delta_2)$ proves a complete class result $\Phi_{q;\psi}(\delta) \geq \Phi_{q;\psi}(\delta_1 \otimes \delta_2)$. Therefore, it remains to find the best product design.

The functions of the Φ_q-criteria for $\psi = \binom{\psi^{(1)}}{\psi^{(2)}}$ split up into their marginal counterparts according to

$$\begin{array}{lcll} \Phi_{q;\psi}(\delta_1 \otimes \delta_2) & = & \Phi_{q;\psi_1}(\delta_1) + \Phi_{q;\psi_2}(\delta_2) & \text{for } 0 < q < \infty, \\ \Phi_{0;\psi}(\delta_1 \otimes \delta_2) & = & \Phi_{0;\psi_1}(\delta_1)\Phi_{0;\psi_2}(\delta_2) & \text{for the } D\text{-criterion } (q = 0), \text{ and} \\ \Phi_{\infty;\psi}(\delta_1 \otimes \delta_2) & = & \max(\Phi_{\infty;\psi_1}(\delta_1), \Phi_{\infty;\psi_2}(\delta_2)) & \text{for the } E\text{-criterion } (q = \infty), \end{array}$$

for every product design $\delta_1 \otimes \delta_2 \in \Delta(\psi)$. Hence, $\Phi_{q;\psi}(\delta_1 \otimes \delta_2)$ attains its minimum if the values $\Phi_{q;\psi_1}(\delta_1)$ and $\Phi_{q;\psi_2}(\delta_2)$ of the marginal criteria are minimized separately. □

In particular, if δ_1^* and δ_2^* are Φ_q-optimum for β_1 and β_2, then $\delta^* = \delta_1^* \otimes \delta_2^*$ is Φ_q-optimum for $\binom{\beta_1}{\beta_2}$. We note also that we can recover Corollary 5.12 by letting $q = 1$. Moreover, Theorem 5.2 concerning the D-optimality with respect to the full parameter vector is a direct consequence of the present Theorem 5.13 in view of Theorem 3.3.

Example 5.11 *K-way layout without interactions (complete control; cf Example 5.5):*

By replacing Corollary 5.12 by Theorem 5.13 in the argumentation of Example 5.10 we obtain a characterization of the D-optimum designs for the parameters associated with the treatment effects of the first M factors:

δ^* is D_ψ-optimum for the effects $(\alpha_{i_k}^{(k)};\ i_k = 1, ..., I_k - 1, k = 1, ..., M)$ of the first M factors compared to control in the K-way layout without interactions if and only if

(i) $\delta_k^*(\{i_k\}) = \frac{1}{I_k}$, $i_k = 1, ..., I_k$, for $k = 1, ..., M$; and

(ii) $\delta_{k_1,k_2}^* = \delta_{k_1}^* \otimes \delta_{k_2}^*$ for $k_1 = 1, ..., M$, $k_2 = k_1 + 1, ...K$.

We, thus, recover Theorem 2 of WIERICH (1986a) in view of Corollary 5.10. □

More generally, we can formulate a result in the spirit of Theorem 4.5 based on the derivative of the criterion function:

Theorem 5.14

A1: *$\psi = (\psi^{(k)})_{k=1,...,K}$, is an s-dimensional minimum linear aspect of β in the additive K-factor model $\mu(t_1, ..., t_K) = \beta_0 + \sum_{k=1}^K f_k(t_k)'\beta_k$, $(t_1, ..., t_K) \in T_1 \times ... \times T_K$, for which the s_k-dimensional components $\psi^{(k)}$ depend on β_k, $\psi^{(k)}(\beta) = L_k\beta_k$, and ψ_k is the associated s_k-dimensional linear aspect of $\beta^{(k)}$, $\psi_k(\beta^{(k)}) = L_k\beta_k$, in the kth marginal model $\mu_k(t_k) = \beta_0 + f_k(t_k)'\beta_k$, $t_k \in T_k$.*

A2: *Let $\varphi : \mathbb{R}^{s\times s} \to \mathbb{R}$ and $\varphi_k : \mathbb{R}^{s_k\times s_k} \to \mathbb{R}$. The criterion function $\Phi : \Delta \to \mathbb{R} \cup \{\infty\}$ is defined by $\Phi(\delta) = \varphi(\mathbf{C}_\psi(\delta))$ for $\delta \in \Delta(\psi)$ and $\Phi(\delta). = \infty$ otherwise, and the marginal crieterion function $\Phi^{(k)} : \Delta_k \to \mathbb{R} \cup \{\infty\}$ is defined by $\Phi^{(k)}(\delta_k) = \varphi_k(\mathbf{C}_{k,\psi_k}(\delta_k))$ for $\delta_k \in \Delta_k(\psi_k)$ and $\Phi^{(k)}(\delta_k) = \infty$ otherwise.*

A3: *Φ and $\Phi^{(k)}$ are convex on Δ and Δ_k, respectively.*

A4: *For δ_k^* the gradient matrices ∇_{φ_k} at $\mathbf{C}_{k,\psi_k}(\delta_k^*)$ and ∇_φ at $\mathbf{C}_\psi(\bigotimes_{k=1}^K \delta_k^*)$ exist and the following representation holds*

$$\nabla_\varphi(\mathbf{C}_\psi(\textstyle\bigotimes_{k=1}^K \delta_k^*)) = \mathrm{diag}((\nabla_{\varphi_k}(\mathbf{C}_{k,\psi_k}(\delta_k^*)))_{k=1,...,K}) .$$

A5: *$\mathbf{I}_k(\delta_k^*)$ is regular, $k = 1, ..., K$.*

If the design δ_k^* is $\Phi^{(k)}$-optimum in the kth marginal model and if the assumptions A1 to A5 hold, then the K-fold product design $\delta^* = \bigotimes_{k=1}^K \delta_k^*$ is Φ-optimum.

Proof

By induction we obtain from Lemma 5.7 that $\delta^* \in \Delta(\beta)$ and from Lemma 5.8 that $\mathbf{C}_\psi(\delta^*) = \mathrm{diag}((\mathbf{C}_{k,\psi_k}(\delta_k^*))_{k=1,...,K})$. Hence, we get

$$L_\psi \mathbf{C}(\delta^*)a(t) = (L_k \mathbf{C}_{k,\beta_k}(\delta_k^*)(f_k(t_k) - \textstyle\int f_k\, d\delta_k^*))_{k=1,...,K} = (L_{k,\psi_k}\mathbf{C}_k(\delta_k^*)a^{(k)}(t_k))_{k=1,...,K}$$

and, consequently,

$$\begin{aligned} & a(t)'\mathbf{C}(\delta^*)L_\psi'\nabla_\varphi(\mathbf{C}_\psi(\delta^*))L_\psi\mathbf{C}(\delta^*)a(t) \\ = & \sum_{k=1}^K a^{(k)}(t_k)'\mathbf{C}_k(\delta_k^*)L_{k,\psi_k}'\nabla_{\varphi_k}(\mathbf{C}_{k,\psi_k}(\delta_k^*))L_{k,\psi_k}\mathbf{C}_k(\delta_k^*)a^{(k)}(t_k) \\ \le & \sum_{k=1}^K \mathrm{tr}(\mathbf{C}_{k,\psi_k}(\delta_k^*)\nabla_{\varphi_k}(\mathbf{C}_{k,\psi_k}(\delta_k^*))) \\ = & \mathrm{tr}(\mathbf{C}_\psi(\delta^*)\nabla_\varphi(\mathbf{C}_\psi(\delta^*))) , \end{aligned}$$

for every $t = (t_1, ..., t_K)$, where the inequality follows from the optimality of δ_k^* by the general equivalence theorem (Theorem 2.5). Thus, the product design δ^* is Φ-optimum in view of the same equivalence theorem. □

The condition A4 on the structure of the gradient matrices is fulfilled for example by the Φ_q-criteria considered in Theorem 5.13, $0 \le q < \infty$.

By using the equivalence theorem for possibly singular information matrices (Theorem 2.10) we can generalize the result of Theorem 5.14, at least, by admitting one of the marginal information matrices $\mathbf{I}_k(\delta_k^*)$ to be singular. Until now it is still an open problem whether condition A5 can be completely removed.

We close this subsection with some investigations which include inference on the constant term β_0. The related results are less satisfactory, however. First we consider the c-criterion which measures the quality of inference on a one-dimensional linear aspect ψ given by $\psi(\beta) = c'\beta$.

Let the vector $c = (c_0, c_1', c_2')'$ of interest be partitioned according to the influence of the constant term and the single factors respectively, $c_0 \in \mathbb{R}$, $c_k \in \mathbb{R}^{p_k - 1}$, and the associated marginal linear aspects $\psi_k(\beta^{(k)}) = c^{(k)\prime}\beta^{(k)}$ are defined by $c^{(k)} = (c_0, c_k')'$. By Lemma 5.8 the variance for $c'\beta$ splits into the marginal variances for $c^{(k)\prime}\beta^{(k)}$ according to $\mathbf{C}_{c'\beta}(\delta_1 \otimes \delta_2) = \mathbf{C}_{1,c^{(1)\prime}\beta^{(1)}}(\delta_1) + \mathbf{C}_{2,c^{(2)\prime}\beta^{(2)}}(\delta_2) - c_0^2$ for every product design $\delta_1 \otimes \delta_2 \in \Delta(c'\beta)$. Thus $\delta_1^* \otimes \delta_2^*$ is the best product design with respect to the c-criterion if δ_1^* and δ_2^* are $c^{(1)}$- resp. $c^{(2)}$-optimum in the corresponding marginal models (compare Theorem 5.6 on A-optimality). The statement that $\delta_1^* \otimes \delta_2^*$ is also c-optimum in $\Delta(c'\beta)$ is only true for such c for which, at least, one component $c^{(k)}$ is a convex combination of the image of the corresponding regression function $a^{(k)}$, up to a multiplicative constant γ, i. e. $\gamma c^{(k)} = \sum_{j=1}^J w_j a^{(k)}(t_k^{(j)})$ for some $t_k^{(j)} \in T_k$, $w_j > 0$, $\sum_{j=1}^J w_j = 1$,

Theorem 5.15

Let $c = (1, c_1', c_2')'$, $c^{(k)} = (1, c_k')'$ *and* c_2 *in the convex hull of* $\{f_2(t_2);\ t_2 \in T_2\}$. *If* δ_k^* *is* $c^{(k)}$*-optimum in the single factor model* $\mu_k(t_k) = \beta_0 + f_k(t_k)'\beta_k$, $t_k \in T_k$, *then the product design* $\delta^* = \delta_1^* \otimes \delta_2^*$ *is c-optimum in the additive model* $\mu(t_1, t_2) = \beta_0 + f_1(t_1)'\beta_1 + f_2(t_2)'\beta_2$, $(t_1, t_2) \in T_1 \times T_2$.

Proof

By a refinement argument we have $\mathbf{C}_{c'\beta}(\delta) \geq \mathbf{C}_{1,c^{(1)\prime}\beta^{(1)}}(\delta_1) \geq \mathbf{C}_{1,c^{(1)\prime}\beta^{(1)}}(\delta_1^*)$ for every design $\delta \in \Delta(c'\beta)$ where δ_1 denotes its first marginal (cf Lemma 3.5). Moreover, we notice that $\delta_1^* \otimes \delta_2^* \in \Delta(c'\beta)$ by Lemma 5.7 and $\mathbf{C}_{2,c^{(2)\prime}\beta^{(2)}}(\delta_2^*) = 1$ in view of Corollary 2.14. Hence, $\mathbf{C}_{c'\beta}(\delta_1^* \otimes \delta_2^*) = \mathbf{C}_{1,c^{(1)\prime}\beta^{(1)}}(\delta_1^*)$ which completes the proof. □

Besides others this result covers the problem of an optimum design for the prediction of the response function $\mu(t^{(0)}) = a(t^{(0)})'\beta$ at a given point $t^{(0)}$ when, at least, one of its components $t_k^{(0)}$ lies within the corresponding design region T_k. As a direct consequence we obtain the following result on optimality for β_0.

Corollary 5.16

If δ_k^* *is* β_0*-optimum in the single factor model* $\mu_k(t_k) = \beta_0 + f_k(t_k)'\beta_k$, $t_k \in T_k$, *and* $\mathbf{0}$ *is in the convex hull of, at least, one of the sets* $\{f_k(t_k); t_k \in T_k\}$, *then the product design* $\delta^* = \delta_1^* \otimes \delta_2^*$ *is* β_0*-optimum in the additive model* $\mu(t_1, t_2) = \beta_0 + f_1(t_1)'\beta_1 + f_2(t_2)'\beta_2$, $(t_1, t_2) \in T_1 \times T_2$.

Generalizations to more than two dimensions are straightforward if we assume that, at least, $K - 1$ components c_k of $c = (1, c_1', ..., c_K')'$ are in the corresponding convex hulls of $\{f_k(t_k); t_k \in T_k\}$. That these conditions on c cannot be omitted is shown by the following example.

Example 5.12 *Two-way layout without interactions (complete control; cf Example 5.1):*

For notational simplicity we consider again the case $I = J = 2$ (cf Examples 5.7 and 5.8). We are interested in the joint effect $c'\beta = \alpha^{(1)} + \alpha^{(2)}$, where $\alpha^{(k)} = \alpha_1^{(k)}$, of the treatments compared to control, i. e. $c = (0, 1, 1)'$. Obviously, the conditions of Theorem 5.15 are not fulfilled because of $c_0 = 0$. By ELFVING's theorem (Theorem 2.13) it can be seen that the design δ^* which assigns equal weights $\frac{1}{2}$ to the design points $(1, 1)$ and $(2, 2)$ is c-optimum. This design δ^* results in a numerical value 4 for $\mathbf{C}_{c'\beta}(\delta^*)$ for $c'\beta$ which is half of the variance $\mathbf{C}_{c'\beta}(\delta_1^* \otimes \delta_2^*) = 2\mathbf{C}_{c^{(1)\prime}\beta^{(1)}}(\delta_1^*) = 8$ obtained for the best product design $\delta_1^* \otimes \delta_2^*$ which assigns equal weights $\frac{1}{4}$ to each of the level combinations. We note that this substantial reduction is caused by the fact that the observations at the design point $(1, 2)$ and $(2, 1)$ are not used for making inference on the aspect $\alpha^{(1)} + \alpha^{(2)}$ if the product design $\delta_1^* \otimes \delta_2^*$ is chosen. We add that this result is not restricted to the case $c_0 = 0$. □

A positive result, however, can be obtained for D-optimality if we are interested in all parameters associated with the effects of one factor and some parameters associated with the other factor, possibly including the constant term,

Theorem 5.17

Let $\psi = \binom{\psi^{(1)}}{\psi^{(2)}}$ *be a linear aspect with components* $\psi^{(1)}(\beta) = \beta_1$ *and* $\psi^{(2)}(\beta) = \psi_2(\beta^{(2)})$. *If* δ_1^* *is D-optimum in the first marginal model* $\mu_1(t_1) = \beta_0 + f_1(t_1)'\beta_1$, $t_1 \in T_1$, *and* δ_2^*

is D_{ψ_2}-optimum in the second marginal model $\mu_2(t_2) = \beta_0 + f_2(t_2)'\beta_2$, $t_2 \in T_2$, then the product design $\delta^* = \delta_1^* \otimes \delta_2^*$ is D_ψ-optimum in the additive model $\mu(t_1, t_2) = \beta_0 + f_1(t_1)'\beta_1 + f_2(t_2)'\beta_2$, $(t_1, t_2) \in T_1 \times T_2$.

Proof

First we notice that $\delta^* \in \Delta(\psi)$ by Lemma 5.7. If we consider the second marginal model as a submodel of the two-factor model without interactions, we obtain $\det(\mathbf{C}_\psi(\delta)) = \det(\mathbf{C}_{\beta_1}(\delta))\det(\mathbf{C}_{2,\psi_2}(\delta_2))$ for every $\delta \in \Delta(\psi)$ by Lemma 3.9 and, hence, $\det(\mathbf{C}_\psi(\delta)) \geq \det(\mathbf{C}_{1,\beta_1}(\delta_1))\det(\mathbf{C}_{2,\psi_2}(\delta_2))$, by a refinement argument (Lemma 3.5), where δ_1 and δ_2 are the marginals of δ. According to Lemma 5.5 equality holds for product designs $\delta = \delta_1 \otimes \delta_2$ which shows that every design is dominated by the product of its marginals. The best product design is given by $\delta^* = \delta_1^* \otimes \delta_2^*$. □

In particular, if δ_1^* is D-optimum and δ_2^* is β_0-optimum, then the product design $\delta^* = \delta_1^* \otimes \delta_2^*$ is D-optimum for $\binom{\beta_0}{\beta_1}$. This most important particular case has been treated by SCHWABE and WIERICH (1995). Because a reparametrization of a subset of the parameters does not affect the D-optimality for those parameters we can obtain an analogous result in the model (5.4) (cf Corollary 5.10) which generalizes part b) of the main Theorem in WIERICH (1989a),

Corollary 5.18

Let δ_1^* be D-optimum in the first marginal model $\mu_1(t_1) = \beta_0 + f_1(t_1)'\beta_1$, $t_1 \in T_1$, and let δ_2^* be β_0-optimum in the second marginal model $\mu_2(t_2) = a^{(2)}(t_2)'\beta_2$, $t_2 \in T_2$. If $\mathbf{1} \in \mathrm{span}(a_1^{(1)}, ..., a_{p_1}^{(1)})$, then the product design $\delta^* = \delta_1^* \otimes \delta_2^*$ is D_{β_1}-optimum in the two-factor model $\mu(t_1, t_2) = a^{(1)}(t_1)'\beta_1 + f_2(t_2)'\beta_2$, $(t_1, t_2) \in T_1 \times T_2$, without interactions.

Example 5.13 *One-way layout with additional regression (cf Example 5.3):*

Assume that we are interested in the direct effects of the different levels $i = 1, ..., I$ for the qualitative factor. Then it is more convenient to use the standard parametrization $\mu(i, u) = \alpha_i + f(u)'\beta_2$, where the parameters α_i are now the direct effects. The first marginal model is a one-way layout (in standard parametrization) and the uniform design δ_1^* is D-optimum which assigns equal weights $\frac{1}{I}$ to each level $i = 1, ..., I$. The second factor may be regarded as a noise factor. In case of an additional linear regression in u, i.e. $\mu_2(u) = \beta_0 + \beta_2 u$, for a β_0-optimum design δ_2^* either the one-point design ϵ_0 at 0 may be used, in case $0 \in U$, or the optimum extrapolation design of Example 2.1 has to be taken. As a constant term is implicitly involved in the one-way layout the resulting product design $\delta_1^* \otimes \delta_2^*$ is D-optimum for the parameter vector $(\alpha_i)_{i=1,...,I}$ of direct effects in the two-factor model. The same design is also D-optimum in the related intra-class model with identical partial models in which there are interactions between the factors (see Example 4.8). □

Concluding remark. In additive models optimum design can be generated as products of those marginals which are optimum in the corresponding marginal models, if the parameters of interest are associated with the effects of the single factors. For the D-criterion the optimality can be extended to the whole aprameter vector including the constant term. Results of this type have been obtained by KUROTSCHKA (1971–1988), GAFFKE and KRAFFT (1979), GAFFKE (1981), PUKELSHEIM (1983), KUROTSCHKA and WIERICH

(1984) and WIERICH (1984–1989), besides others, for a variety of particular models. In case of a given marginal design the D-optimality of product designs was obtained by COOK and THIBODEAU (1980). The general approach treated here was developed by SCHWABE (1995b) and, for the D-optimality, by SCHWABE and WIERICH (1995). Some efficiency bounds for product designs were calculated by WIERICH (1989b) and SCHWABE and WONG (1995).

5.2 Orthogonal Designs

In this subsection we discuss the construction of optimum designs as product designs in situations where a constant term is not necessarily involved, but where a strong interplay of optimum marginal designs and marginal regression functions is inherent.

In this setting the K-factor models without interactions are given by

$$\mu(t_1, ..., t_K) = a(t_1, ..., t_K)'\beta = \sum_{k=1}^{K} a^{(k)}(t_k)'\beta_k \ , \tag{5.5}$$

$t = (t_1, ..., t_K) \in T = T_1 \times ... \times T_K$, such that $a(t_1, ..., t_K) = (a^{(1)}(t_1)', ..., a^{(K)}(t_K)')'$, $\beta_k \in \mathbb{R}^{p_k}$ and $\beta = (\beta_k)_{k=1,...,K} \in \mathbb{R}^p$ with $p = \sum_{k=1}^{K} p_k$.

We notice that model (5.1) which explicitly involves a constant term can always be rewritten into the form (5.5). However, the corresponding marginals will be different from (5.2) as we consider the marginal models

$$\mu_k(t_k) = a^{(k)}(t_k)'\beta_k \ , \tag{5.6}$$

$t_k \in T_k$, $k = 1, ..., K$, with no constant term explicitly involved in the present situation. Indeed, we assume that all regression functions in the K-factor model are linearly independent and a constant term may occur in at most one of the marginal models. Without loss of generality this constant term will appear in the first marginal model in case a constant term is present in the whole model. Note that, thus, in general, the marginal models considered here are different from the conditional models as introduced in the introduction of this chapter.

As before we will be mainly interested in the two-factor model without interactions

$$\mu(t_1, t_2) = a^{(1)}(t_1)'\beta_1 + a^{(2)}(t_2)'\beta_2 \ , \tag{5.7}$$

$(t_1, t_2) \in T_1 \times T_2$, such that $a(t_1, t_2) = \binom{a^{(1)}(t_1)}{a^{(2)}(t_2)}$ and $\beta = \binom{\beta_1}{\beta_2} \in \mathbb{R}^p$ with $p = p_1 + p_2$. However, all results can be extended to K factors by induction.

The product of optimum designs will be optimum in the present setting if we accept the restrictive additional assumption that the marginal regression functions $a^{(k)}$ are *centered* with respect to the optimum marginal design δ_k^* in, at least, all but one marginal models, i. e.

$$\int a^{(k)} \, d\delta_k^* = \mathbf{0} \tag{5.8}$$

for $k = 2, ..., K$ (see also KUROTSCHKA (1988)). If we consider the property (5.8) as a condition on the design, we will say that a design δ_k is *orthogonal* if $\int a^{(k)} \, d\delta_k = \mathbf{0}$. Of course, the first marginal regression function $a^{(1)}$ cannot be centered if a constant term is present in the model.

An important advantage of orthogonal designs δ_k is the resulting orthogonality of any two regression functions belonging to different factors with respect to the K-fold product design $\delta = \bigotimes_{k=1}^K \delta_k$ (see Lemma 5.20). Hence, the estimators for the parameters belonging to different factors will be uncorrelated and the covariance matrix is block diagonal with its marginal counterparts represented in the blocks.

As in the previous subsection we start with a preparatory result on the connection between identifiability in the two-factor model and identifiability in the marginal models,

Lemma 5.19

Let ψ be a linear aspect of β, $\psi(\beta) = L_\psi \beta \in \mathbb{R}^s$, let $L_\psi = (L_1|L_2)$ be partitioned according to the influence of the single factors on ψ, i. e. $L_k \in \mathbb{R}^{s \times p_k}$, and let ψ_k be the associated linear aspect in the marginal model, $\psi_k(\beta_k) = L_k\beta_k$.

(i) *If $\delta_1 \in \Delta_1(\psi_1)$ and $\delta_2 \in \Delta_2(\psi_2)$ then $\delta_1 \otimes \delta_2 \in \Delta(\psi)$.*

(ii) *If $\delta \in \Delta(\psi)$ then $\delta_k \in \Delta_k(\psi_k)$ and, consequently, $\delta_1 \otimes \delta_2 \in \Delta(\psi)$.*

(iii) *Let ψ be a linear aspect of β_k only, $\psi(\beta) = \psi_k(\beta_k)$. Then $\delta_k \in \Delta_k(\psi_k)$ implies $\delta_1 \otimes \delta_2 \in \Delta(\psi)$ for every marginal design δ_2 on T_2 (for $k = 1$) resp. for every marginal design δ_1 on T_1 ($k = 2$).*

Proof

The proof parallels that of Lemma 5.7 with obvious modifications. □

In what follows the main tool is again a representation of the generalized inverse of the information matrix $\mathbf{I}(\delta_1 \otimes \delta_2)$ of a product design $\delta_1 \otimes \delta_2$ in the two-factor model without interactions expressed in terms of the corresponding marginal information matrices,

Lemma 5.20

If $\int a^{(2)}\, d\delta_2 = \mathbf{0}$, then the matrix

$$\mathbf{I}(\delta_1 \otimes \delta_2)^- = \begin{pmatrix} \mathbf{I}_1(\delta_1)^- & \mathbf{0} \\ \mathbf{0} & \mathbf{I}_2(\delta_2)^- \end{pmatrix}$$

is a generalized inverse of the information matrix $\mathbf{I}(\delta_1 \otimes \delta_2)$, where $\mathbf{I}_k(\delta_k)$ is the information matrix in the kth marginal model for the marginal design δ_k.

Proof

As $\int a^{(1)} a^{(2)\prime}\, d(\delta_1 \otimes \delta_2) = \mathbf{0}$ the information matrix

$$\mathbf{I}(\delta_1 \otimes \delta_2) = \begin{pmatrix} \mathbf{I}_1(\delta_1) & \mathbf{0} \\ \mathbf{0} & \mathbf{I}_2(\delta_2) \end{pmatrix}$$

has a block diagonal structure, and the result is straightforward. □

We can now present a powerful complete class theorem in analogy to Theorem 5.9,

Theorem 5.21

Let $\psi^{(k)}$ be a linear aspect depending on the kth partial parameter vector β_k, $\psi^{(k)}(\beta) = \psi_k(\beta_k)$, in the two-factor model $\mu(t_1, t_2) = a^{(1)}(t_1)'\beta_1 + a^{(2)}(t_2)'\beta_2$, $(t_1, t_2) \in T_1 \times T_2$, without interactions.

(i) *If $\delta \in \Delta(\psi^{(2)})$ and $a^{(2)}$ is centered with respect to δ_2, then $\mathbf{C}_{\psi^{(2)}}(\delta) \geq \mathbf{C}_{\psi^{(2)}}(\delta_1 \otimes \delta_2)$, where δ_1 and δ_2 are the marginals of δ.*

(ii) *If there exists an orthogonal design $\widetilde{\delta}_2$ on T_2, then $\mathbf{C}_{\psi^{(1)}}(\delta) \geq \mathbf{C}_{\psi^{(1)}}(\delta_1 \otimes \widetilde{\delta}_2)$, for every $\delta \in \Delta(\psi^{(1)})$, where δ_1 is the first marginal of δ.*

Proof

We use Lemmas 5.19 and 5.20 instead of Lemmas 5.7 and 5.8 and replace J_1 and J_2 by $\mathbf{I}_1$ and $\mathbf{I}_2$, respectively, in the proof of Theorem 5.9. □

By a refinement argument we obtain for the covariance matrices $\mathbf{C}_{\beta_k}(\delta) \geq \mathbf{C}_{k,\beta_k}(\delta_k)$ for $\delta \in \Delta(\beta_k)$ with equality if and only if $a^{(1)}$ and $a^{(2)}$ are *orthogonal* with respect to δ, i. e. $\int a^{(1)}a^{(2)\prime}\, d\delta = \mathbf{0}$. In case of equality we obtain, additionally, $\mathbf{C}_{\beta_k}(\delta) = \mathbf{C}_{\beta_k}(\delta_1 \otimes \delta_2)$.

From Theorem 5.21 we can directly derive that product designs are A-opimum for linear aspects $\psi = \binom{\psi^{(1)}}{\psi^{(2)}}$ if their components $\psi^{(k)}$ depend only on the partial parameter vectors β_k and the regression functions are centered with respect to the optimum marginal design in the second marginal model (cf Corollary 5.12),

Corollary 5.22

Let $\psi = \binom{\psi^{(1)}}{\psi^{(2)}}$ be given by $\psi^{(k)}(\beta) = \psi_k(\beta_k)$. If δ_k^* is A_{ψ_k}-optimum in the single factor model $\mu_k(t_k) = a^{(k)}(t_k)'\beta_k$, $t_k \in T_k$, and if $\int a^{(2)}\, d\delta_2^* = \mathbf{0}$, then the product design $\delta^* = \delta_1^* \otimes \delta_2^*$ is A_ψ-optimum in the two-factor model $\mu(t_1, t_2) = a^{(1)}(t_1)'\beta_1 + a^{(2)}(t_2)'\beta$, $(t_1, t_2) \in T_1 \times T_2$, without interactions.

Proof

By Theorem 5.21 $\delta_1^* \otimes \delta_2^*$ is $A_{\psi^{(1)}}$- and $A_{\psi^{(2)}}$-optimum. The relation $\mathrm{tr}(\mathbf{C}_\psi(\delta)) = \mathrm{tr}(\mathbf{C}_{\psi^{(1)}}(\delta)) + \mathrm{tr}(\mathbf{C}_{\psi^{(2)}}(\delta))$ for every $\delta \in \Delta(\psi)$ establishes the A_ψ-optimality. □

In particular, if δ_1^* and δ_2^* are A-optimum in the marginal models and $\int a^{(2)}\, d\delta_2^* = \mathbf{0}$, then the product design $\delta^* = \delta_1^* \otimes \delta_2^*$ is A-optimum.

We note that in contrast to Example 5.7 we can verify the A-optimality of product designs for the full parameter vector in the two-factor model without interactions, if the regression functions are centered with respect to the A-optimum marginal design for (at least) one factor,

Example 5.14 *One-way layout with additional linear regression (cf Example 5.3):*

As in Example 5.13 we consider the more natural standard parametrization (without general mean) for the one-way layout part $\mu(i, u) = \alpha_i + \beta_2 u$ assuming that the control variable u for the straight line regression part may be taken from a symmetric interval $U = [-u_0, u_0]$. The first marginal model is a one-way layout $\mu_1(i) = \alpha_i$, in standard parametrization, and the second marginal model is a linear regression without intercept, $\mu_2(u) = \beta_2 u$, $u \in [-u_0, u_0]$, and, hence, differs from the representation given in Examples 5.3 and 5.13 because the constant term is suppressed.

The marginal design δ_1^* which assigns equal weights $\frac{1}{I}$ to each level is A- (and D-) optimum (see Example 3.4). In the second marginal model the information matrix reduces to the real-valued expression $\int u^2\, \delta_2(du)$ which attains its maximum u_0^2 for those optimum designs $\widetilde{\delta}_2^{(w)}$ which assign weight w to the right endpoint u_0 and weight $1 - w$ to the left endpoint $-u_0$ of the symmetric interval, $0 \leq w \leq 1$. Because of $\int u\, \widetilde{\delta}_2^{(w)}(du) = (2w - 1)u_0$ the design $\delta_2^* = \delta_2^{(w^*)}$ with $w^* = \frac{1}{2}$ is the unique orthogonal optimum design and, hence, the product design $\delta_1^* \otimes \delta_2^*$ is A-optimum.

After a reparametrization to the representation given in Example 5.3 it can be shown by Theorem 5.2 that the same product design $\delta_1^* \otimes \delta_2^*$ is also D-optimum. □

The result of Corollary 5.22 can be generalized to all criteria based on the eigenvalues,

Theorem 5.23

Let $\psi = \binom{\psi^{(1)}}{\psi^{(2)}}$ be given by $\psi^{(k)}(\beta) = \psi_k(\beta_k)$. If δ_k^* is Φ_q-optimum for ψ_k in the single factor model $\mu_k(t_k) = a^{(k)}(t_k)'\beta_k$, $t_k \in T_k$, and if $\int a^{(2)}\, d\delta_2^* = \mathbf{0}$, then the product design $\delta^* = \delta_1^* \otimes \delta_2^*$ is Φ_q-optimum for ψ in the two-factor model $\mu(t_1, t_2) = a^{(1)}(t_1)'\beta_1 + a^{(2)}(t_2)'\beta_2$, $(t_1, t_2) \in T_1 \times T_2$, without interactions.

Proof

By the formula for a generalized inverse of a partitioned matrix (see Lemma A.3) the matrix

$$\mathbf{I}(\delta)^- = \begin{pmatrix} \mathbf{I}_1^- + \mathbf{I}_1^-\mathbf{I}_{12}(\mathbf{I}_2 - \mathbf{I}_{12}'\mathbf{I}_1^-\mathbf{I}_{12})^-\mathbf{I}_{12}'\mathbf{I}_1^- & -\mathbf{I}_1^-\mathbf{I}_{12}(\mathbf{I}_2 - \mathbf{I}_{12}'\mathbf{I}_1^-\mathbf{I}_{12})^- \\ -(\mathbf{I}_2 - \mathbf{I}_{12}'\mathbf{I}_1^-\mathbf{I}_{12})^-\mathbf{I}_{12}'\mathbf{I}_1^- & (\mathbf{I}_2 - \mathbf{I}_{12}'\mathbf{I}_1^-\mathbf{I}_{12})^- \end{pmatrix}$$

is a generalized inverse of the information matrix $\mathbf{I}(\delta)$, where $\mathbf{I}_{12} = \int a^{(1)}a^{(2)\prime}\, d\delta$ and $\mathbf{I}_k = \mathbf{I}_k(\delta_k)$ is the information matrix in the kth marginal model for the marginal design δ_k of δ. With this representation and with $J(\delta)$, J_1, J_2 and J_{12} replaced by $\mathbf{I}(\delta)$, $\mathbf{I}_1$, $\mathbf{I}_2$ and $\mathbf{I}_{12}$, respectively, the proof follows along the lines indicated in the proof of Theorem 5.13 by formal sign change. □

In particular, if δ_1^* and δ_2^* are Φ_q-optimum in the marginal models and $\int a^{(2)}\, d\delta_2^* = \mathbf{0}$, then the product design $\delta^* = \delta_1^* \otimes \delta_2^*$ is Φ_q-optimum.

By letting $q = 1$ we recover the A-optimality result of Corollary 5.22, and for the limiting cases $q = 0$ and $q = \infty$ results on D- and E-optimality are obtained.

Example 5.15 *Chemical balance weighing:*

We consider the problem of weighing K objects on an unbiased chemical balance. If we denote the weight of the kth object by β_k, the experiment can be described by

$$\mu(t_1, ..., t_K) = \sum_{k=1}^K t_k\beta_k\ ,$$

$t_k \in \{-1, 0, 1\}$, $k = 1, ..., K$, where the levels have the following interpretation: t_k equals 1 if the kth object is on the left pan, t_k equals -1 if the kth object is on the right pan, and t_k equals 0 if the kth object is not present in the weighing arrangement. Hence, the response μ is the difference of the total weight of the objects collected in the left pan minus the total weight of the objects in the right pan which can be observed at the scale subject to a random error. No constant term occurs since the balance is assumed to be unbiased. The marginal models are all identical $\mu_k(t_k) = t_k\beta_k$, $t_k \in \{-1, 0, 1\}$, and represent the experimental situation for weighing one object on a chemical balance.

The marginal problem can completely be solved: Any design which assigns weight zero to the design point 0, i. e. for which the object is permamently present in the weighing arrangements, is optimum. Moreover, the optimum design δ_k^* which assigns equal weight $\frac{1}{2}$ to $+1$ and -1 is orthogonal (cf Example 5.14).

Hence, by an iterated application of Theorem 5.23 the product design $\delta^* = \bigotimes_{k=1}^K \delta_k^*$ is A-, D-, and E-optimum which assigns equal weight 2^{-K} to each of those weighing arrangement in which all K objects are present (cf RAGHAVARAO (1971)). We note that one of the marginal designs δ_k^*, say δ_1^*, may be replaced by an arbitrary optimum marginal design $\widetilde{\delta}_1$ (i. e. $\widetilde{\delta}_1(\{0\}) = 0$) without changing the optimality.

The product design $\delta^* = \bigotimes_{k=1}^K \delta_k^*$ in which all marginals are orthogonal remains A-, D-, and E-optimum if the balance may be biased and an additional constant term is introduced (cf CHENG (1988)). □

As indicated in the above example the result of Theorem 5.23 can be extended to K factors, $K \geq 2$, by induction if the regression functions are centered with respect to the optimum marginal designs for at least $K-1$ factors involved. This assertion will be expanded in the following generalization of a theorem by RAFAJŁOWICZ and MYSZKA (1992) to potentially singular information matrices (cf Theorem 4.5):

Theorem 5.24

A1: *$\psi = (\psi^{(k)})_{k=1,...,K}$ is an s-dimensional minimum linear aspect of β in the K-factor model $\mu(t_1,...,t_K) = \sum_{k=1}^K a^{(k)}(t_k)'\beta_k$, $(t_1,...,t_K) \in T_1\times...\times T_K$, without interactions for which the s_k-dimensional components $\psi^{(k)}$ depend on β_k, $\psi^{(k)}(\beta) = \psi_k(\beta_k)$, and ψ_k is the associated linear aspect of β_k in the marginal model $\mu_k(t_k) = a^{(k)}(t_k)\beta_k$, $t_k \in T_k$.*

A2: *Let $\varphi : I\!R^{s\times s} \to I\!R$ and $\varphi_k : I\!R^{s_k\times s_k} \to I\!R$. The criterion function $\Phi : \Delta \to I\!R \cup \{\infty\}$ is defined by $\Phi(\delta) = \varphi(\mathbf{C}_\psi(\delta))$ for $\delta \in \Delta(\psi)$ and $\Phi(\delta) = \infty$ otherwise, and the marginal criterion function $\Phi^{(k)} : \Delta_k \to I\!R \cup \{\infty\}$ is defined by $\Phi^{(k)}(\delta_k) = \varphi_k(\mathbf{C}_{k,\psi_k}(\delta_k))$ for $\delta_k \in \Delta_k(\psi_k)$ and $\Phi^{(k)}(\delta_k) = \infty$ otherwise.*

A3: *Φ and $\Phi^{(k)}$ are convex on Δ and Δ_k, respectively.*

A4: *For δ_k^* the gradient matrices ∇_{φ_k} at $\mathbf{C}_{k,\psi_k}(\delta_k^*)$ and ∇_φ at $\mathbf{C}_\psi(\bigotimes_{k=1}^K \delta_k^*)$ exist and the following representation holds*

$$\nabla_\varphi(\mathbf{C}_\psi(\bigotimes_{k=1}^K \delta_k^*)) = \mathrm{diag}((\nabla_{\varphi_k}(\mathbf{C}_{k,\psi_k}(\delta_k^*)))_{k=1,...,K}) \; .$$

A5: *$\int a^{(k)}\, d\delta_k^* = \mathbf{0}$ for $k = 2,...,K$.*

If the designs δ_k^* are $\Phi^{(k)}$-optimum in the kth marginal model and if the assumptions A1 to A5 hold, then the K-fold product design $\delta^* = \bigotimes_{k=1}^K \delta_k^*$ is Φ-optimum.

Proof

By induction we obtain from Lemma 5.19 that $\delta^* \in \Delta(\psi)$ and from Lemma 5.20 that $\mathbf{C}_\psi(\delta^*) = \mathrm{diag}((\mathbf{C}_{k,\psi_k}(\delta_k^*))_{k=1,...,K})$. The equivalence theorem for possibly singular information matrices (Theorem 2.10) states that there exist matrices H_k such that $\mathrm{rank}(H_k) = p_k - \mathrm{rank}(\mathbf{I}_k(\delta_k^*))$, $\mathbf{I}_k(\delta_k^*) + H_k'H_k$ is regular, and for every $t_k \in T_k$

$$\begin{aligned} & a^{(k)}(t_k)'(\mathbf{I}_k(\delta_k^*) + H_k'H_k)^{-1}L_{k,\psi_k}'\nabla_{\varphi_k}(\mathbf{C}_{k,\psi_k}(\delta_k^*))L_{k,\psi_k}(\mathbf{I}_k(\delta_k^*) + H_k'H_k)^{-1}a^{(k)}(t_k) \\ \leq \;& \mathrm{tr}(\mathbf{C}_{k,\psi_k}(\delta_k^*)\nabla_{\varphi_k}(\mathbf{C}_{k,\psi_k}(\delta_k^*))) \; . \end{aligned}$$

Define $H = \mathrm{diag}((H_k)_{k=1,...,K})$, then $\mathbf{I}(\delta^*) + H'H = \mathrm{diag}((\mathbf{I}_k(\delta_k^*) + H_k'H_k)_{k=1,...,K})$ is regular and $\mathrm{rank}(H) = p - \mathrm{rank}(\mathbf{I}(\delta^*))$. All matrices are block diagonal and, thus,

$$\begin{aligned} & a(t)'(\mathbf{I}(\delta^*) + H'H)^{-1}L_\psi'\nabla_\varphi(\mathbf{C}_\psi(\delta^*))L_\psi(\mathbf{I}(\delta^*) + H'H)^{-1}a(t) \\ = \;& \sum_{k=1}^K a^{(k)}(t_k)'(\mathbf{I}_k(\delta_k^*) + H_k'H_k)^{-1}L_{k,\psi_k}'\nabla_{\varphi_k}(\mathbf{C}_{k,\psi_k}(\delta_k^*))L_{k,\psi_k}(\mathbf{I}_k(\delta_k^*) + H_k'H_k)^{-1}a^{(k)}(t_k) \\ \leq \;& \sum_{k=1}^K \mathrm{tr}(\mathbf{C}_{k,\psi_k}(\delta_k^*)\nabla_{\varphi_k}(\mathbf{C}_{k,\psi_k}(\delta_k^*))) \\ = \;& \mathrm{tr}(\mathbf{C}_\psi(\delta^*)\nabla_\varphi(\mathbf{C}_\psi(\delta^*))) \end{aligned}$$

for every $t = (t_1, ..., t_K) \in T_1 \times ... \times T_K$. Hence, the design δ^* is Φ-optimum by Theorem 2.10. □

Note that the Φ_q-optimality results of Theorem 5.23 are included, for $q < \infty$. However, neither c-optimality nor Q-optimality are covered, in general, and counterexamples can be constructed similarly to those of Examples 5.7 and 5.12.

To illustrate the fact that the restrictions of the present section cannot be dropped we start with an example which shows that in the absence of a constant term product designs can be far from being optimum.

Example 5.16 *Spring balance weighing:*

As in Example 5.15 we consider the problem of weighing K objects, but now on an unbiased spring balance. The parameters β_k denote again the weights of the kth object, and the experiment will be modeled by

$$\mu(t_1, ..., t_K) = \sum_{k=1}^{K} t_k\beta_k$$

as in Example 5.15 for the chemical balance, but with a different underlying design region $T_k = \{0, 1\}$, $k = 1, ..., K$. Here t_k equals one or zero corresponding to whether the kth object lies on the pan or not. As the balance is assumed to be unbiased no constant term is present. The marginal models are all identical $\mu_k(t_k) = t_k\beta_k$, $t_k \in \{0, 1\}$, and represent the experimental situation for weighing one object on a spring balance.

For the marginal problem we notice that β_k is one-dimensional and, hence, all optimality criteria coincide. The optimum design assigns all mass to 1, i. e. the object should always be put into the pan. This is not surprising, because a weighing arrangement with no object on the pan produces no information on the weights if the balance is known to be unbiased. The resulting product design $\bigotimes_{k=1}^{K} \epsilon_1$ assigns all mass to $(1, ..., 1)$, i. e. to the weighing arrangement in which all objects are present in the pan. This procedure will be optimum if we are interested in the total weight $\sum_{k=1}^{K} \beta_k$, but all other linear aspects are not identifiable and the product design cannot be used.

As a compromise one may look for the best product design. Now, every product design (different from $\bigotimes_{k=1}^{K} \epsilon_1$) assigns positive weight to the design point $(0, ..., 0)$ where nothing but the random noise can be observed because $\mu(0, ..., 0)$ is known to be zero. So the product design can be uniformly improved by distributing the weight assigned to the degenerate experimental condition $(0, ..., 0)$ among the weighing arrangements with just one object included.

For illustrative purposes we examine the case of two objects to be weighed on a spring balance $\mu(t_1, t_2) = t_1\beta_1 + t_2\beta_2$, $t_1, t_2 \in \{0, 1\}$. By considerations of invariance with respect to permutations of the factors we may restrict to designs with equal marginals $\delta_1 = \delta_2$ (cf the remark following Example 3.4). The best product design with respect to the D-criterion is given by the marginals $\delta_k(\{1\}) = \frac{1}{\sqrt{2}}$ and $\delta_k(\{0\}) = \frac{1}{\sqrt{2}}(\sqrt{2} - 1)$ and results in a determinant $\det(\mathbf{C}(\delta_1 \otimes \delta_2)) = 4$ of the covariance matrix $\mathbf{C}(\delta_1 \otimes \delta_2)$. Now by distributing the weight assigned to $(0, 0)$ among $(1, 0)$ and $(0, 1)$ we obtain an improved design $\widetilde{\delta}$ with $\widetilde{\delta}(\{(1, 1)\}) = \frac{1}{2}$, $\widetilde{\delta}(\{(1, 0)\}) = \widetilde{\delta}(\{(0, 1)\}) = \frac{1}{4}$, and the determinant $\det(\mathbf{C}(\widetilde{\delta})) = 3.2$ is considerably smaller. The design δ^* which assigns equal proportions $\frac{1}{3}$ to the three relevant settings $(1, 0)$, $(0, 1)$, and $(1, 1)$, is seen to be D-optimum by the KIEFER-WOLFOWITZ equivalence theorem (Theorem 2.1) and results in a minimum value $\det(\mathbf{C}(\delta^*)) = 3$ for the determinant.

In the general case of K objects there are D-optimum designs which are uniform on a comparatively small number of supporting points (cf HUDA and MUKERJEE (1988)). □

Concluding remark. In models without interactions optimum design can be generated as products of those marginals which are optimum in the corresponding marginal models, if an additional orthogonality property is inherent with respect to those optimum marginals. The advantage of the orthogonal structure was pointed out by KUROTSCHKA (1984) and it has been applied to a variety of models and criteria by RAFAJŁOWICZ and MYSZKA (1992). A more general treatment is indicated in SCHWABE (1995b).

6 Partial Interactions

In the previous Sections 4 and 5 we have treated experimental situations either with complete product-type interactions or with no interactions and discovered that the same product designs are D-optimum in both cases. This result can be extended to the intermediate interaction structure of complete M-factor interactions which will be shown in Subsection 6.1. In the second subsection we treat the frequently occuring situation that the model associated with one of the factors is invariant with respect to a group of transformations which acts transitively on the marginal design region. In this case the restriction to product designs is possible due to the invariance structure. However, in contrast to the previous results the marginals of the optimum designs are not necessarily optimum in the marginal models.

6.1 Complete M-factor Interactions

Throughout this subsection we consider a K-factor model with all those interactions present in which up to M factors are involved,

$$\begin{aligned}\mu(t_1,...,t_K) &= a(t_1,...,t_K)'\beta \\ &= \beta_0 + \sum_{k=1}^{K} f_k(t_k)'\beta_k \\ &\quad + \sum_{k_1=1}^{K}\sum_{k_2=k_1+1}^{K}(f_{k_1}(t_{k_1})\otimes f_{k_2}(t_{k_2}))'\beta_{k_1,k_2} \\ &\quad \vdots \\ &\quad + \sum_{1\le k_1<k_2<...<k_M\le K}(f_{k_1}(t_{k_1})\otimes f_{k_2}(t_{k_2})\otimes ...\otimes f_{k_M}(t_{k_M}))'\beta_{k_1,k_2,...,k_M} \\ &= \beta_0 + \sum_{m=1}^{M}\sum_{1\le k_1<...<k_m\le K}(\otimes_{\ell=1}^{m} f_{k_\ell}(t_{k_\ell}))'\beta_{k_1,...,k_m}\,, \end{aligned} \tag{6.1}$$

$(t_1,...,t_K)\in T_1\times...\times T_k$, which will be called a *model with complete M-factor interactions.*

The corresponding marginal models are given by

$$\mu_k(t_k) = a^{(k)}(t_k)'\beta^{(k)} = \beta_0 + f_k(t_k)'\beta_k\,, \tag{6.2}$$

$t_k\in T_k$, $\beta^{(k)}\in \mathbb{R}^{p_k}$, $k=1,...,K$, and they contain a constant term each as was the case for additive models treated in Subsection 5.1. Hence for $M=1$ there are no interaction terms and the model (6.1) coincides with the K-factor version of the additive model (5.1). For $M=K$ we obtain a model with complete product-type interactions and (6.1) is a special case of the K-factor model considered in Section 4 with a constant term present in all marginal models. The scope of the models included is illustrated by the following examples,

Example 6.1 *K-way layout with complete M-factor interactions:*

$$\begin{aligned}\mu(i_1,...,i_K) &= \mu_0 + \sum_{k=1}^{K}\alpha_{i_k}^{(k)} + \sum_{k_1=1}^{K}\sum_{k_2=k_1+1}^{K}\alpha_{i_{k_1},i_{k_2}}^{(k_1,k_2)} \\ &\quad + ... + \sum_{1\le k_1<k_2<...<k_M\le K}\alpha_{i_{k_1},i_{k_2},...,i_{k_M}}^{(k_1,k_2,...,k_M)} \\ &= \mu_0 + \sum_{m=1}^{M}\sum_{1\le k_1<...<k_m\le K}\alpha_{i_{k_1},...,i_{k_m}}^{(k_1,...,k_m)}\,,\end{aligned}$$

$i_k = 1, ..., I_k$, $k = 1, ..., K$. As there are too many parameters we have to impose identifiability conditions (cf Example 5.5 for the particular case $M = 1$ of non-interacting factors). We consider the parametrization of *complete control* as in Example 5.5, i.e. for each factor k there is a control level, say I_k. This is accomplished by the following identifiability conditions

$$\begin{aligned}
\alpha^{(k)}_{I_k} &= 0 \quad \text{for } k = 1, ..., K, \\
\alpha^{(k_1,k_2)}_{I_{k_1},i_{k_2}} &= \alpha^{(k_1,k_2)}_{i_{k_1},I_{k_2}} = 0 \quad \text{for } i_{k_1} = 1, ..., I_{k_1},\ i_{k_2} = 1, ..., I_{k_2},\ 1 \leq k_1 < k_2 \leq K \\
&\vdots \\
\alpha^{(k_1,k_2,...,k_M)}_{I_{k_1},i_{k_2},...,i_{k_M}} &= \alpha^{(k_1,k_2,...,k_M)}_{i_{k_1},I_{k_2},...,i_{k_M}} = ... = \alpha^{(k_1,k_2,...,k_M)}_{i_{k_1},i_{k_2},...,I_{k_M}} = 0 \\
&\text{for } i_{k_\ell} = 1, ..., I_{k_\ell},\ \ell = 1, ..., M,\ 1 \leq k_1 < k_2 < ... < k_M \leq K.
\end{aligned}$$

Now, with the regression functions f_k given by $f_k(i_k) = \mathbf{e}_{I_k-1,i_k}$ which were introduced in the marginal models of a one-way layout with control (see Example 3.4) the present K-way layout is a model with complete M-factor interactions. □

Example 6.2 *K-dimensional linear regression with M-factor interactions:*

$$\begin{aligned}
\mu(t_1, ..., t_K) &= \beta_0 + \sum_{k=1}^K \beta_k t_k + \sum_{k_1=1}^K \sum_{k_2=k_1+1}^K \beta_{k_1,k_2} t_{k_1} t_{k_2} \\
&\quad + ... + \sum_{1 \leq k_1 < k_2 < ... < k_M \leq K} \beta_{k_1,k_2,...,k_M} t_{k_1} t_{k_2} \cdot ... \cdot t_{k_M} ,
\end{aligned}$$

$(t_1, ..., t_K) \in T_1 \times ... \times T_K \subseteq \mathbb{R}^K$. The marginal models are described by linear regressions,

$$\mu_k(t_k) = \beta_0 + \beta_k t_k ,$$

$t_k \in T_k$, each. □

The first result of this section is concerned with optimum designs in case the highest interaction terms are of interest. We define the collection of the parameters of the m-factor interactions $\beta_{k_1,...,k_m}$ as

$$\beta_{(m)} = (\beta_{k_1,...,k_m})_{1 \leq k_1 < ... < k_m \leq K} \in \mathbb{R}^{p_{(m)}}$$

listed in lexicographic order, $p_{(m)} = \sum_{1 \leq k_1 < ... < k_m \leq K} \prod_{\ell=1}^m (p_{k_\ell} - 1)$ and, hence, $\beta = (\beta_0, \beta'_{(1)}, ..., \beta'_{(M)})'$. The regression functions $a = (1, a'_{(1)}, ..., a'_{(M)})'$ are partitioned appropriately, where $a_{(m)}(t_1, ..., t_K) = (\bigotimes_{\ell=1}^m f_{k_\ell}(t_{k_\ell}))_{1 \leq k_1 < ... < k_m \leq K}$ is the vector of regression function associated with the m-factor interactions. Thus, the underlying model can be rewritten as

$$\mu(t_1, ..., t_K) = \beta_0 + \sum_{m=1}^M a_{(m)}(t_1, ..., t_K)' \beta_{(m)} .$$

As in Section 5 we can simplify the computation of the covariance matrix for a given product design $\delta = \bigotimes_{k=1}^K \delta_k$ if we reparametrize the whole model by switching to the centered versions $\widetilde{f}_k = f_k - \int f_k \, d\delta_k$ of the regression functions in the marginal models,

$$\mu(t_1, ..., t_K) = \widetilde{\beta}_0 + \sum_{m=1}^M \widetilde{a}_{(m)}(t_1, ..., t_K)' \widetilde{\beta}_{(m)}$$

with $\widetilde{a}_{(m)}(t_1, ..., t_K) = (\bigotimes_{\ell=1}^{m} \widetilde{f}_{k_\ell}(t_{k_\ell}))_{1 \le k_1 < ... < k_m \le K}$.

The transformation matrix $\widetilde{L}$ which gives $\widetilde{a} = \widetilde{L}a$ for the new set of regression functions $\widetilde{a} = (1, \widetilde{a}'_{(1)}, ..., \widetilde{a}'_{(M)})'$ is lower triangular with identity matrices of appropriate size $p_{(m)}$ on its diagonal. In particular, the highest interaction terms $\beta_{(M)} = \widetilde{\beta}_{(M)}$ remain unchanged by this transformation $\beta = \widetilde{L}'\widetilde{\beta}$.

Lemma 6.1

Let $\delta = \bigotimes_{k=1}^{K} \delta_k$, $\widetilde{\mathbf{I}}(\delta) = \int \widetilde{a}\widetilde{a}'\, d\delta$, *and* $\widetilde{\mathbf{I}}_{(m)}(\delta) = \int \widetilde{a}_{(m)}\widetilde{a}_{(m)}\, d\delta$, $m = 1, ..., M$.

(i) *The matrices* $\widetilde{\mathbf{I}}(\delta)$ *and* $\widetilde{\mathbf{I}}_{(m)}(\delta)$ *are block diagonal,*

$$\widetilde{\mathbf{I}}_{(m)}(\delta) = \mathrm{diag}((\bigotimes_{\ell=1}^{m} \int \widetilde{f}_{k_\ell}\widetilde{f}'_{k_\ell}\, d\delta_{k_\ell})_{1 \le k_1 < ... < k_m \le K})$$

$$\widetilde{\mathbf{I}}(\delta) = \begin{pmatrix} 1 & \mathbf{0} \\ \mathbf{0} & \mathrm{diag}((\widetilde{\mathbf{I}}_{(m)}(\delta))_{m=1,...,M}) \end{pmatrix} .$$

(ii) $\delta \in \Delta(\beta)$ *if and only if* $\delta \in \Delta(\beta_{(M)})$ *if and only if* $\delta_k \in \Delta_k(\beta_k)$, $k = 1, ..., K$.

(iii) *If* $\delta \in \Delta(\beta)$, *then* $\mathbf{C}_{\beta_{(M)}}(\delta) = \mathrm{diag}((\mathbf{C}_{\beta_{k_1,...,k_M}}(\delta))_{1 \le k_1 < ... < k_M \le K})$.

Proof

(i) The transformed partial information matrix $\widetilde{\mathbf{I}}_{(m)}(\delta)$ consists of blocks of the form $\int (\bigotimes_{\ell=1}^{m} \widetilde{f}_{k_\ell})(\bigotimes_{\ell=1}^{m} \widetilde{f}_{\kappa_\ell})'\, d\delta$, $1 \le k_1 < ... < k_m \le K$, $1 \le \kappa_1 < ... < \kappa_m \le K$. For a given block let $g_{1k} = \widetilde{f}_k$ for $k \in \{k_1, ..., k_m\}, g_{1k} = 1$ otherwise, and $g_{2k} = \widetilde{f}_k$ for $k \in \{\kappa_1, ..., \kappa_m\}, g_{2k} = 1$ otherwise. Then $\bigotimes_{\ell=1}^{m} \widetilde{f}_{k_\ell} = \bigotimes_{k=1}^{K} g_{1k}$ and $\bigotimes_{\ell=1}^{m} \widetilde{f}_{\kappa_\ell} = \bigotimes_{k=1}^{K} g_{2k}$ respectively. As $\delta = \bigotimes_{k=1}^{K} \delta_k$ is a product design integration and Kronecker product may be interchanged in an appropriate way

$$\int (\bigotimes_{\ell=1}^{m} \widetilde{f}_{k_\ell})(\bigotimes_{\ell=1}^{m} \widetilde{f}_{\kappa_\ell})'\, d\delta = \int (\bigotimes_{k=1}^{K} g_{1k}g'_{2k})\, d(\bigotimes_{k=1}^{K} \delta_k) = \bigotimes_{k=1}^{K} \int g_{1k}g'_{2k}\, d\delta_k .$$

For every off-diagonal block there exists an index $k_\ell \notin \{\kappa_1, ..., \kappa_m\}$. Thus, $\int g_{1k_\ell}g'_{2k_\ell}\, d\delta_{k_\ell} = \int \widetilde{f}_{k_\ell} d\delta_{k_\ell} = \mathbf{0}$ and the whole block is a zero matrix. In a diagonal block we have $k_\ell = \kappa_\ell$ for all $\ell = 1, ..., m$ and, hence, $\int g_{1k}g'_{2k}\, d\delta_k = \int \widetilde{f}_k\widetilde{f}'_k\, d\delta_k$ for $k \in \{k_1, ..., k_m\}$ and $\int g_{1k}g'_{2k}\, d\delta_k = 1$ otherwise. This proves the representation of $\widetilde{\mathbf{I}}_{(m)}(\delta)$.

With an obvious modification of the arguments used above we can show $\int \widetilde{a}_{(m)}\widetilde{a}'_{(\tilde{m})} d\delta = \mathbf{0}$ for $m \ne \tilde{m}$ which yields the representation of $\widetilde{\mathbf{I}}(\delta)$.

(ii) By (i) the matrices $\widetilde{\mathbf{I}}(\delta)$ and $\widetilde{\mathbf{I}}_{(M)}(\delta)$ are regular if and only if all of their blocks $\int \widetilde{f}_k\widetilde{f}'_k\, d\delta_k$ on the diagonal are regular. The regularity of $\widetilde{\mathbf{I}}(\delta)$ and $\widetilde{\mathbf{I}}_{(M)}(\delta)$ is equivalent to the identifiability of β and $\beta_{(M)} = \widetilde{\beta}_{(M)}$ respectively, whereas $\int \widetilde{f}_k\widetilde{f}'_k\, d\delta_k$ is regular if and only if β_k is identifiable in the marginal model.

(iii) By part (i) and (ii) the covariance matrix for the transformed parameter vector $\widetilde{\beta}$ is given by $\mathbf{C}_{\widetilde{\beta}}(\delta) = \widetilde{\mathbf{I}}(\delta)^{-1}$ and the result follows from the fact that the parameters $\beta_{k_1,...,k_M} = \widetilde{\beta}_{k_1,...,k_M}$ and, hence, $\beta_{(M)} = \widetilde{\beta}_{(M)}$ associated with the highest order interactions are left unchanged under this transformation. □

Using the representation of Lemma 6.1 we can deal with the highest order interaction terms as if the marginal regression functions were mutually uncorrelated in case we consider a product design.

Theorem 6.2

If δ_k^* is A- (resp. D-) optimum for β_k in the single factor model $\mu_k(t_k) = \beta_0 + f_k(t_k)'\beta_k$, $t_k \in T_k$, then the product design $\delta^* = \bigotimes_{k=1}^K \delta_k^*$ is A- (resp. D-) optimum for the parameter vector $\beta_{(M)}$ of the highest interaction terms in the K-factor model with complete M-factor interactions $\mu(t_1, ..., t_K) = \beta_0 + \sum_{m=1}^M \sum_{1 \le k_1 < ... < k_m \le K} (\bigotimes_{\ell=1}^m f_{k_\ell}(t_{k_\ell}))'\beta_{k_1,...,k_m}$, $(t_1, ..., t_K) \in T_1 \times ... \times T_K$.

Proof

The identifiability of $\beta_{(M)}$ under δ^* is guaranteed by Lemma 6.1 (ii). We introduce the M-dimensional marginal models

$$\begin{aligned} \mu_{k_1,...,k_M}(t_{k_1}, ..., t_{k_M}) \;=\; \beta_0 \;&+\; \sum\nolimits_{\kappa \in \{k_1,...,k_M\}} f_\kappa(t_\kappa)'\beta_\kappa \\ &+\; \sum\nolimits_{\kappa_1,\kappa_2 \in \{k_1,...,k_M\}, \kappa_1 < \kappa_2} (f_{\kappa_1}(t_{\kappa_1}) \otimes f_{\kappa_2}(t_{\kappa_2}))'\,\beta_{\kappa_1,\kappa_2} \\ &\;\;\vdots \\ &+\; \left(\bigotimes\nolimits_{\ell=1}^M f_{k_\ell}(t_{k_\ell})\right)'\beta_{k_1,k_2,...,k_M}\ , \end{aligned}$$

$t_{k_\ell} \in T_{k_\ell}$, $\ell = 1, ..., M$, $1 \le k_1 < k_2 < ... < k_M \le K$, which are submodels of the whole model (6.1) and can be written in terms of the regression functions $a^{(k)} = (1, f_k')'$ for the marginal models as $\mu_{k_1,...,k_M}(t_{k_1}, ..., t_{k_M}) = (\bigotimes_{\ell=1}^M a^{(k_\ell)}(t_{k_\ell}))'\beta^{(k_1,...,k_M)}$. Hence, these M-dimensional submodels have complete product-type interactions and can be treated by the methods of Section 4. Moreover, by the refinement argument of Lemma 3.5 we obtain $\mathbf{C}_{\beta_{k_1,...,k_M}}(\delta) \ge \mathbf{C}_{k_1,...,k_M,\beta_{k_1,...,k_M}}(\delta_{k_1,...,k_M})$, $1 \le k_1 < ... < k_M \le K$, where $\mathbf{C}_{k_1,...,k_M,\beta_{k_1,...,k_M}}(\delta_{k_1,...,k_M})$ is the covariance matrix for $\beta_{k_1,...,k_M}$ in the M-dimensional marginal model with factors $k_1, ..., k_M$ when the underlying design is the corresponding M-dimensional marginal $\delta_{k_1,...,k_M}$ of δ.

(i) *A-criterion*

By applying Corollary 4.6 to the M-dimensional submodel associated with the factors $k_1, ..., k_M$ we see (by letting $\psi_\ell = \beta_{k_\ell}$) that $\mathrm{tr}(\mathbf{C}_{k_1,...,k_M,\beta_{k_1,...,k_M}}(\delta_{k_1,...,k_M}))$ attains its minimum for $\delta_{k_1,...,k_M}^* = \bigotimes_{\ell=1}^M \delta_{k_\ell}^*$. Hence, by Lemma 6.1

$$\begin{aligned} \mathrm{tr}(\mathbf{C}_{\beta_{(M)}}(\delta)) &= \sum\nolimits_{1 \le k_1 < ... < k_M \le K} \mathrm{tr}(\mathbf{C}_{\beta_{k_1,...,k_M}}(\delta)) \\ &\ge \sum\nolimits_{1 \le k_1 < ... < k_M \le K} \mathrm{tr}(\mathbf{C}_{k_1,...,k_M,\beta_{k_1,...,k_M}}(\delta_{k_1,...,k_M})) \\ &\ge \sum\nolimits_{1 \le k_1 < ... < k_M \le K} \mathrm{tr}(\mathbf{C}_{k_1,...,k_M,\beta_{k_1,...,k_M}}(\bigotimes\nolimits_{\ell=1}^M \delta_{k_\ell}^*)) \\ &= \sum\nolimits_{1 \le k_1 < ... < k_M \le K} \mathrm{tr}(\mathbf{C}_{\beta_{k_1,...,k_M}}(\bigotimes\nolimits_{k=1}^K \delta_k^*)) \\ &= \mathrm{tr}(\mathbf{C}_{\beta_{(M)}}(\delta^*))\ . \end{aligned}$$

(ii) *D-criterion*

We directly apply the equivalence theorem for D_ψ-optimality (Corollary 2.6) to the present situation. As the covariance matrix $\mathbf{C}(\delta^*)$ is related to the transformed information matrix $\widetilde{\mathbf{I}}(\delta^*)$ by $\mathbf{C}(\delta^*) = \widetilde{L}'\widetilde{\mathbf{I}}(\delta^*)^{-1}\widetilde{L}$ we obtain $L_{\beta_{(M)}}\mathbf{C}(\delta^*)a(t) = \mathbf{C}_{\beta_{(M)}}(\delta^*)\widetilde{a}_{(M)}(t)$ and, similarly, $L_{\beta_{k_1,...,k_M}}\mathbf{C}(\delta^*)a(t) = \mathbf{C}_{k_1,...,k_M}(\bigotimes_{\ell=1}^M \delta_{k_\ell}^*)(\bigotimes_{\ell=1}^M \widetilde{f}_{k_\ell}(t_{k_\ell}))$. By Corollary 4.6 the product design $\bigotimes_{\ell=1}^M \delta_{k_\ell}^*$ is D-optimum for $\beta_{k_1,...,k_M}$ in the associated M-dimensional submodel. Thus, the equivalence theorem for D_ψ-optimality (Corollary 2.6) applied to every M-dimensional submodel yields

$$\begin{aligned} & a(t)'\mathbf{C}(\delta^*)L'_{\beta_{(M)}}\mathbf{C}_{\beta_{(M)}}(\delta^*)^{-1}L_{\beta_{(M)}}\mathbf{C}(\delta^*)a(t) \\ = & \sum\nolimits_{1\le k_1<\ldots<k_M\le K}(\bigotimes_{\ell=1}^M \tilde{f}_{k_\ell}(t_{k_\ell}))'\mathbf{C}_{k_1,\ldots,k_M,\beta_{k_1,\ldots,k_M}}(\bigotimes_{\ell=1}^M \delta^*_{k_\ell})(\bigotimes_{\ell=1}^M \tilde{f}_{k_\ell}(t_{k_\ell})) \\ \le & \; p_{(M)} \end{aligned}$$

which proves the assertion, again, by the equivalence theorem. □

Both methods of proof used in Theorem 6.2 - orthogonalization followed by a complete class argument and mechanical application of the equivalence theorem - can be extended to generalize the theorem to other criteria, e. g. to the Φ_q-criteria, $0 \le q < \infty$. With respect to the D-optimality a more general result can be shown,

Theorem 6.3

If δ_k^* is D-optimum in the single factor model $\mu_k(t_k) = \beta_0 + f_k(t_k)'\beta_k$, $t_k \in T_k$, then $\delta^* = \bigotimes_{k=1}^K \delta_k^*$ is D-optimum for the parameter vector $\beta_{(M_1,\ldots,M)} = (\beta'_{(M_1)},\ldots,\beta'_{(M)})'$ of the $M - M_1 + 1$ highest order interaction terms in the K-factor model with complete M-factor interactions $\mu(t_1,\ldots,t_K) = \beta_0 + \sum_{m=1}^M \sum_{1\le k_1<\ldots<k_m\le K}(\bigotimes_{\ell=1}^m f_{k_\ell}(t_{k_\ell}))'\beta_{k_1,\ldots,k_m}$, $(t_1,\ldots,t_K) \in T_1 \times \ldots \times T_K$, for every $M_1 = 1,\ldots,M$.

Moreover, $\delta^* = \bigotimes_{k=1}^K \delta_k^*$ is D-optimum for the whole parameter vector β.

Proof

We note that δ_k^* is D-optimum for β_k in the kth marginal model in view of Theorem 3.3 and that $\delta^* \in \Delta(\beta)$ according to Lemma 6.1.

For $M_1 = M$ the D-optimality has been shown in Theorem 6.2. We now perform an induction backward to 1: Assume the design δ^* is D-optimum for $\beta_{(M_1+1,\ldots,M)}$. By Lemma 3.9 we have $\det(\mathbf{C}_{\beta_{(M_1,\ldots,M)}}(\delta)) = \det(\mathbf{C}_{\beta_{(M_1+1,\ldots,M)}}(\delta))\det(\mathbf{C}^{(M_1)}_{\beta_{(M_1)}}(\delta))$ where $\mathbf{C}^{(M_1)}_{\beta_{(M_1)}}(\delta)$ is the covariance matrix for the vector $\beta_{(M_1)}$ of the highest order interaction terms in the K-factor model with complete M_1-factor interactions. In the last equation both terms on the right hand side attain their minimum for $\delta = \delta^*$, the second by Theorem 6.2 and the first by the assumption made in the induction, which proves the assertion. In particular, we have shown that δ^* is D-optimum for $\beta_{(1,\ldots,M)}$ which in view of Theorem 3.3 yields the D-optimality of the design δ^* for the whole parameter vector β. □

Theorem 6.3 indicates that the product design $\delta^* = \bigotimes_{k=1}^K \delta_k^*$ is D-optimum irrespectively of the actual degree M of the interactions. Moreover, the same design is also D-optimum for discriminating between models of different degree M_1 and M_2. In this sense the product design $\delta^* = \bigotimes_{k=1}^K \delta_k^*$ is robust against a misspecification of the model.

We mention that in view of Theorems 6.2 and 6.3 it is sufficient for δ^* to be optimum that in case $M < K/2$ the $(2M)$-dimensional marginals of δ^* are *proportional*, i. e. $\delta_{k_1,\ldots,k_{2M}} = \bigotimes_{\ell=1}^{2M} \delta_{k_\ell}$ for every $1 \le k_1 < \ldots k_{2M} \le K$, because only those marginals are involved in the information matrix. In many situation designs with this proportionality property are known to exist with a much smaller number of supporting points than the product design, e.g. fractional factorials, orthogonal arrays etc. (cf the remark following Example 5.5).

Example 6.3 *K-way layout with complete M-factor interactions (cf Example 6.1):* The design δ^* which assigns equal weight $\prod_{k=1}^{K} \frac{1}{I_k}$ to each level combination is D-optimum for the whole parameter vector β and for the parameter vectors $\beta_{(M_1,\dots,M)}$ of the $M - M_1 + 1$ highest order interaction terms, $M_1 = 1, \dots, M$.

This is not true for the smallest interaction terms or e.g. the main effects $\beta_{(1)}$ which will be illustrated by the example of a two-way layout $\mu(i,j) = \mu_0 + \alpha_i^{(1)} + \alpha_j^{(2)} + \alpha_{ij}^{(1,2)}$, with complete interactions ($M = K = 2$). Here the identifiability conditions $\alpha_I^{(1)} = \alpha_J^{(2)} = \alpha_{Ij}^{(1,2)} = \alpha_{iJ}^{(1,2)} = 0$ of complete control are imposed. We can rewrite the model in standard parametrization $\mu(i,j) = \alpha_{ij}$ and see from the transformation matrix that the constant term μ_0 and the main effects $\beta_{(1)} = (\alpha_1^{(1)}, \dots, \alpha_{I-1}^{(1)}, \alpha_1^{(2)}, \dots, \alpha_{J-1}^{(2)})$ depend only on the parameters $(\alpha_{I1}, \dots, \alpha_{IJ}, \alpha_{1J}, \dots, \alpha_{(I-1)J})$, i. e. the inference is based only on those level combinations in which, at least, one control level is present. Hence, we can restrict the design region to those level combinations, $T^\nabla = \{(i,j) \in T;\, i = I \text{ or } j = J\}$, and the designs supported on T^∇ obviously constitute an essentially complete class in Δ. On T^∇ the model reduces to a two-way layout without interactions $\mu(i,j) = \mu_0 + \alpha_i^{(1)} + \alpha_j^{(2)}$ on a reduced design region $(i,j) \in T^\nabla$, with identifiability conditions $\alpha_I^{(1)} = \alpha_J^{(2)} = 0$ and the meaning of the parameters is preserved from the original model. We notice that the number $I + J - 1$ of parameters equals the number of design points in T^∇. Hence, by Lemma 2.11 the uniform design δ^∇ which assigns equal weights $\frac{1}{I+J-1}$ to every design point in T^∇ is D-optimum for $\beta_{(0,1)}$. By Theorem 3.3 the same design δ^∇ is also D-optimum for the main effects $\beta_{(1)}$. By the complete class argument the optimality results of δ^∇ carry over to the original model. □

In the presence of orthogonal optimum marginal designs further results can be obtained in the spirit of Theorem 5.24 with the methods employed here. In particular, for the linear regression model (Example 6.2) with complete M-factor interactions on the cube $[-1,1]^K$ the product design $\delta^* = \bigotimes_{k=1}^{K} \delta_k^*$ which assigns equal weights 2^{-K} to each corner $(t_1, \dots, t_K) \in \{-1,1\}^K$ of the cube is optimum with respect to every Φ_q-criterion which includes A-, D-, and E-optimality.

Concluding remark. In models with complete M-factor interactions optimum design for the highet interaction terms can be generated as products of those marginals which are optimum in the corresponding marginal models. Moreover, the product of D-optimum marginals results in a design which is D-optimum for the whole parameter vector independent of the actual depth of interactions and, additionally, D-optimum for comparing different orders of interactions. In the particular model of a K-way layout KUROTSCHKA (1972) raised the question for an optimum design which was solved via an approach to general models by SCHWABE (1995a). Extensions to other criteria in case of an underlying orthogonal structure are given in SCHWABE (1995c) (cf RICCOMAGNO, SCHWABE and WYNN (1995). and for more general complete interaction structures see SCHWABE (1995d).

6.2 Invariant Designs

This subsection is devoted to experimental situations in which all levels of one factor are of the same influence. In most cases this factor will be qualitative, i. e. it may be adjusted to a finite number of levels, and permutations, i. e. rearrangements of the ordering of the levels, do not affect the performance of the experiment. These requirements have to be made both on the model and on the criterion function of interest. The formalization will be done in terms of induced linear transformations and invariance as introduced in Subsection 3.2.

In general, we consider the two-factor model

$$\mu(t_1, t_2) = (a^{(1)}(t_1) \otimes f_1(t_2))'\beta_1 + f_2(t_2)'\beta_2 , \tag{6.3}$$

$(t_1, t_2) \in T_1 \times T_2$. The first marginal model

$$\mu_1(t_1) = a^{(1)}(t_1)'\beta^{(1)} , \tag{6.4}$$

$t_1 \in T_1$, with $\beta^{(1)} \in \mathbb{R}^{p_1}$, describes the factor for which the levels are of equal influence. To be more precise we assume that a group G_1 of transformations on the marginal design region T_1 induces linear transformations of the marginal regression function $a^{(1)}$. This marginal group G_1 induces a group $G = \{g;\, g : (t_1, t_2) \to (g_1(t_1), t_2), g_1 \in G_1\}$ of transformations on the common design region $T_1 \times T_2$ by $g(t_1, t_2) = (g_1(t_1), t_2)$, for $g_1 \in G_1$, and G induces linear transformations of the regression function $a = ((a^{(1)} \otimes f_1)', f_2')'$ of the whole model,

$$a(g(t_1, t_2)) = \begin{pmatrix} a^{(1)}(g_1(t_1)) \otimes f_1(t_2) \\ f_2(t_2) \end{pmatrix} = \begin{pmatrix} (Q_{1,g_1} a^{(1)}(t_1)) \otimes f_1(t_2) \\ f_2(t_2) \end{pmatrix} = Q_g a(t_1, t_2) ,$$

where $Q_{1,g_1} \in \mathbb{R}^{p_1 \times p_1}$ and

$$Q_g = \begin{pmatrix} Q_{1,g_1} \otimes \mathbf{E}_{p_{2,1}} & \mathbf{0} \\ \mathbf{0} & \mathbf{E}_{p_{2,2}} \end{pmatrix} \in \mathbb{R}^{p \times p}$$

are the transformation matrices corresponding to the group elements $g_1 \in G_1$ and $g = (g_1, \mathrm{id}|_{T_2}) \in G$, respectively. The second marginal model is given by

$$\mu_2(t_2) = f_1(t_2)'\beta_1^{(2)} + f_2(t_2)'\beta_2 , \tag{6.5}$$

$t_2 \in T_2$. Let $p_{2,1}$ and $p_{2,2}$ be the dimensions of the components f_1 and f_2 of the p_2-dimensional regression function $a^{(2)} = (f_1', f_2')'$ in the second marginal model. Note that $p_2 = p_{2,1} + p_{2,2}$ and $p = p_1 p_{2,1} + p_{2,2}$ for the dimensions involved and that the mapping $g_1 \to g = (g_1, \mathrm{id}|_{T_2})$ defines a natural one-to-one relationship between G_1 and G.

As mentioned earlier this situation typically arises for the first factor being qualitative. In particular, many of our examples deal with the special class of models

$$\mu(i, u) = f_1(u)'\beta_{1,i} + f_2(u)'\beta_2 , \tag{6.6}$$

$i = 1, ..., I$, $u \in U$, such that $\beta_1 = (\beta_{1,i})_{i=1,...,I} \in \mathbb{R}^{Ip_{2,1}}$, and the first marginal model is a one-way layout (cf Example 3.3)

$$\mu_1(i) = \alpha_i , \tag{6.7}$$

$i = 1, ..., I$. For the transformations g_1 acting on the design region $\{1, ..., I\}$ we consider the group of permutations which describe a rearrangement or a relabeling of the levels of the first factor. The corresponding transformation matrices Q_{1,g_1} are orthogonal, $Q'_{1,g_1} = Q^{-1}_{1,g_1}$. The orthogonality carries over to Q_g because of

$$Q'_g = \begin{pmatrix} Q'_{1,g_1} \otimes \mathbf{E}_{p_{2,1}} & \mathbf{0} \\ \mathbf{0} & \mathbf{E}_{p_{2,2}} \end{pmatrix} = \begin{pmatrix} Q^{-1}_{1,g_1} \otimes \mathbf{E}_{p_{2,1}} & \mathbf{0} \\ \mathbf{0} & \mathbf{E}_{p_{2,2}} \end{pmatrix} = Q_g^{-1} .$$

Hence all Φ_q-criteria, $0 \leq q \leq \infty$, including the A-, D- and E-criterion are invariant with respect to G in the present model (6.6) (see Theorem 3.12).

Example 6.4 *One-way layout combined with linear regression:*
Additionally to the one-way layout $\mu_1(i) = \alpha_i$, $i = 1, ..., I$, the marginal model for the second factor is specified by

$$\mu_2(u) = \beta_0 + \beta_1 u ,$$

$u \in U$. In this setting of a two-factor model we can consider different types of interaction, in which the slope or the intercept or both may depend on the level i of the first factor.

a) *Complete interactions:*

$$\mu(i, u) = \alpha_i + \beta_{1,i} u .$$

This is the intra-class model treated in Section 4.

b) *Common slope:*

$$\mu(i, u) = \alpha_i + \beta_1 u$$

(see SEARLE (1971), pp 348). This model is obtained by partitioning the regression function $a^{(2)}$ of the second marginal model according to $f_1(u) = 1$ and $f_2(u) = u$. For this *model without interactions* it has been established in Theorem 5.2 that product designs are D-optimum with D-optimum marginals.

c) *Common intercept:*
By interchanging the roles of f_1 and f_2 in b) we obtain

$$\mu(i, u) = \beta_0 + \beta_{1,i} u ,$$

which can be used e.g. to model situations where I different treatments are applied alternatively and the value u of the second factor corresponds to the dose of the treatment chosen; in particular, $u = 0$ corresponds to no treatment (cf Example 7.2 and BUONACCORSI (1986)). □

In the general model (6.3) the first factor will be called *invariant* if the corresponding group G_1 of transformations acts transitively on T_1, i.e. if the orbit $G_1(t_1) = \{g_1(t_1); g_1 \in G_1\}$ of t_1 under G_1 equals the whole design region T_1 for every $t_1 \in T_1$ (see Definiton 3.7). In this situation the uniform design $\overline{\delta}_1$ on T_1 is the unique invariant design with respect to G_1 by Lemma 3.14.

To simplify the forthcoming argumentations we assume, as in Subsection 3.2, that the marginal design region T_1 of the first factor is finite, which is always true for qualitative factors. We note that the finiteness of T_1 implies that G_1 and hence G are also finite and that the uniform design $\overline{\delta}_1$ assigns weights $\frac{1}{\#T_1}$ to each level $t_1 \in T_1$, where $\#T_1$ denotes the number of levels in T_1. (The application to more general concepts will be indicated at the end of this subsection in Example 6.12 in which the invariant factor will be modeled by a trigonometric regression.) The next result establishes a characterization of the invariant designs under the present assumptions,

Lemma 6.4

Let T_1 be finite, $\overline{\delta}_1$ the uniform design on T_1, let G_1 induce linear transformations of $a^{(1)}$ and let G_1 act transitively on T_1. Then a design $\delta \in \Delta$ is invariant with respect to G if and only if $\delta = \overline{\delta}_1 \otimes \delta_2$.

Proof

As in the proof of Lemma 3.14 let $N_1 = \#T_1$ be the number of design points in T_1. For every $t_1 \in T_1$ there are $g_1^{(1)}, ..., g_1^{(N_1)} \in G_1$ such that $\{g_1^{(1)^{-1}}(t_1), ..., g_1^{(N_1)^{-1}}(t_1)\} = T_1$ because G_1 acts transitively on T_1. Denote by $g^{(n)} = (g_1^{(n)}, \mathrm{id}|_{T_2}) \in G$ the corresponding transformations on $T_1 \times T_2$.

If δ is invariant with respect to G then $\delta(\{t_1\} \times B_2) = \frac{1}{N_1} \sum_{n=1}^{N_1} \delta(\{g_1^{(n)^{-1}}(t_1)\} \times B_2) = \overline{\delta}_1(\{t_1\})\delta_2(B_2)$ for every $t_1 \in T_1$ and every (measurable) $B_2 \subseteq T_2$ where δ_2 is the second marginal of δ, hence $\delta = \overline{\delta}_1 \otimes \delta_2$.

For the converse part we notice that for every $g \in G$ there is a $g_1 \in G_1$ with $g = (g_1, \mathrm{id}|_{T_2})$. Thus for every $t_1 \in T_1$ and every (measurable) $B_2 \subseteq T_2$ we get $\delta^g(\{t_1\} \times B_2) = \overline{\delta}_1(\{g_1^{-1}(t_1)\})\delta_2(B_2) = \overline{\delta}_1(\{t_1\})\delta_2(B_2) = \delta(\{t_1\} \times B_2)$. Hence, $\delta^g = \delta$ for every $g \in G$ and δ is invariant. □

Again, we note that under the assumption of a finite support for δ no measure theory is needed in the proof of Lemma 6.4 because only finite subsets B_2 have to be considered. For the invariant designs determined by Lemma 6.4 we are able to give a useful representation of a generalized inverse of the information matrix depending on whether a constant term is involved in the first marginal model or not,

Lemma 6.5

Let T_1 be finite, $\overline{\delta}_1$ the uniform design on T_1, let G_1 induce linear transformations of $a^{(1)}$ and let G_1 act transitively on T_1.

(i) *If $\mathbf{1} \in \mathrm{span}(a_1^{(1)}, ..., a_{p_1}^{(1)})$, then a generalized inverse of the information matrix*

$$\mathbf{I}(\overline{\delta}_1 \otimes \delta_2) = \begin{pmatrix} \int a^{(1)}a^{(1)\prime}\, d\overline{\delta}_1 \otimes \int f_1 f_1'\, d\delta_2 & \int a^{(1)}\, d\overline{\delta}_1 \otimes \int f_1 f_2'\, d\delta_2 \\ (\int a^{(1)}\, d\overline{\delta}_1)' \otimes \int f_2 f_1'\, d\delta_2 & \int f_2 f_2'\, d\delta_2 \end{pmatrix}$$

is given by

$$\mathbf{I}(\overline{\delta}_1 \otimes \delta_2)^- = \begin{pmatrix} \mathbf{I}_1(\overline{\delta}_1)^- \otimes (\int f_1 f_1'\, d\delta_2)^- + \widetilde{L}_1'\widetilde{L}_1 \otimes \widetilde{L}_2' J_{22}^- \widetilde{L}_2 & -\widetilde{L}_1' \otimes \widetilde{L}_2' J_{22}^- \\ -\widetilde{L}_1 \otimes J_{22}^- \widetilde{L}_2 & J_{22}^- \end{pmatrix}$$

where $\mathbf{I}_1(\overline{\delta}_1) = \int a^{(1)}a^{(1)\prime}\, d\overline{\delta}_1$ is the information matrix in the first marginal model and $\widetilde{L}_1$, $\widetilde{L}_2$, and J_{22} are defined by

$$\begin{aligned} \widetilde{L}_1 &= (\textstyle\int a^{(1)}\, d\overline{\delta}_1)'\mathbf{I}_1(\overline{\delta}_1)^- , \\ \widetilde{L}_2 &= \textstyle\int f_2 f_1'\, d\delta_2 (\int f_1 f_1'\, d\delta_2)^- , \\ J_{22} &= \textstyle\int f_2 f_2'\, d\delta_2 - \int f_2 f_1'\, d\delta_2 (\int f_1 f_1'\, d\delta_2)^- \int f_1 f_2'\, d\delta_2 . \end{aligned}$$

(ii) *If $\mathbf{1} \notin \mathrm{span}(a_1^{(1)}, ..., a_{p_1}^{(1)})$, then*

$$\mathbf{I}(\overline{\delta}_1 \otimes \delta_2)^- = \begin{pmatrix} \mathbf{I}_1(\overline{\delta}_1)^- \otimes (\int f_1 f_1'\, d\delta_2)^- & \mathbf{0} \\ \mathbf{0} & (\int f_2 f_2'\, d\delta_2)^- \end{pmatrix}$$

is a generalized inverse of the information matrix

$$\mathbf{I}(\overline{\delta}_1 \otimes \delta_2) = \begin{pmatrix} \int a^{(1)}a^{(1)\prime}\, d\overline{\delta}_1 \otimes \int f_1 f_1'\, d\delta_2 & \mathbf{0} \\ \mathbf{0} & \int f_2 f_2'\, d\delta_2 \end{pmatrix}.$$

Proof

Similar to the proofs of Lemma 5.5 and Lemma 5.8 we use the technique of centering with respect to the design $\delta = \overline{\delta}_1 \otimes \delta_2$. We define the function $\widetilde{f}_2 : T_1 \times T_2 \to \mathbb{R}^{p_{2,2}}$ by $\widetilde{f}_2(t_1, t_2) = f_2(t_2) - \widetilde{L}_1 a^{(1)}(t_1) \otimes \widetilde{L}_2 f_1(t_2)$. Then $a^{(1)} \otimes f_1$ and $\widetilde{f}_2$ are *orthogonal* with respect to $\overline{\delta}_1 \otimes \delta_2$, i. e. $\int (a^{(1)} \otimes f_1)\widetilde{f}_2'\, d(\overline{\delta}_1 \otimes \delta_2) = \mathbf{0}$, and, additionally,

$$\begin{aligned} \int \widetilde{f}_2 \widetilde{f}_2'\, d(\overline{\delta}_1 \otimes \delta_2) &= \int f_2 f_2'\, d\delta_2 - (\widetilde{L}_1 \mathbf{I}_1(\overline{\delta}_1)\widetilde{L}_1') \otimes (\widetilde{L}_2 \int f_1 f_1'\, d\delta_2 \widetilde{L}_2') \\ &= \int f_2 f_2'\, d\delta_2 - (\int a^{(1)}\, d\overline{\delta}_1)'\mathbf{I}_1(\overline{\delta}_1)^- \int a^{(1)}\, d\overline{\delta}_1 \cdot \int f_2 f_1'\, d\delta_2 (\int f_1 f_1'\, d\delta_2)^- \int f_1 f_2'\, d\delta_2 \end{aligned}$$

where the last equality follows from $\int f_1 f_1'\, d\delta_2 (\int f_1 f_1'\, d\delta_2)^- \int f_1 f_2'\, d\delta_2 = \int f_1 f_2'\, d\delta_2$ and $\mathbf{I}_1(\overline{\delta}_1)\mathbf{I}_1(\overline{\delta}_1)^- \int a_1\, d(\overline{\delta}_1) = \int a_1\, d(\overline{\delta}_1)$ (see Lemma 3.1). We define the *centered* regression function $\widetilde{a}$ by $\widetilde{a} = ((a^{(1)} \otimes f_1)', \widetilde{f}_2')'$. Then

$$\widetilde{L} = \begin{pmatrix} \mathbf{E}_{p_1} \otimes \mathbf{E}_{p_{2,1}} & \mathbf{0} \\ -\widetilde{L}_1 \otimes \widetilde{L}_2 & \mathbf{E}_{p_{2,2}} \end{pmatrix}$$

is the transformation matrix for $\widetilde{a} = \widetilde{L}a$ and the transformed information matrix

$$\widetilde{\mathbf{I}}(\overline{\delta}_1 \otimes \delta_2) = \int \widetilde{a}\widetilde{a}'\, d(\overline{\delta}_1 \otimes \delta_2) = \begin{pmatrix} \mathbf{I}_1(\overline{\delta}_1) \otimes \int f_1 f_1'\, d\delta_2 & \mathbf{0} \\ \mathbf{0} & \int \widetilde{f}_2 \widetilde{f}_2'\, d(\overline{\delta}_1 \otimes \delta_2) \end{pmatrix}$$

is block diagonal. A generalized inverse of $\widetilde{\mathbf{I}}(\overline{\delta}_1 \otimes \delta_2)$ is obtained by inverting the blocks separately and, consequently,

$$\begin{aligned} \mathbf{I}(\overline{\delta}_1 \otimes \delta_2)^- &= \widetilde{L}'\widetilde{\mathbf{I}}(\overline{\delta}_1 \otimes \delta_2)^- \widetilde{L} \\ &= \begin{pmatrix} \begin{array}{l} \mathbf{I}_1(\overline{\delta}_1)^- \otimes (\int f_1 f_1'\, d\delta_2)^- \\ \quad + \widetilde{L}_1'\widetilde{L}_1 \otimes \widetilde{L}_2'(\int \widetilde{f}_2 \widetilde{f}_2'\, d(\overline{\delta}_1 \otimes \delta_2))^- \widetilde{L}_2 \end{array} & -\widetilde{L}_1' \otimes \widetilde{L}_2'(\int \widetilde{f}_2 \widetilde{f}_2'\, d(\overline{\delta}_1 \otimes \delta_2))^- \\ -\widetilde{L}_1 \otimes (\int \widetilde{f}_2 \widetilde{f}_2'\, d(\overline{\delta}_1 \otimes \delta_2))^- \widetilde{L}_2 & (\int \widetilde{f}_2 \widetilde{f}_2'\, d(\overline{\delta}_1 \otimes \delta_2))^- \end{pmatrix}. \end{aligned}$$

is a generalized inverse of $\mathbf{I}(\overline{\delta}_1 \otimes \delta_2)$.

If $\mathbf{1} \in \text{span}(a_1^{(1)}, ..., a_{p_1}^{(1)})$, then $(\int a^{(1)}\, d\overline{\delta}_1)'(\int a^{(1)}a^{(1)\prime}\, d\overline{\delta}_1)^- \int a^{(1)}\, d\overline{\delta}_1 = 1$ according to Lemma 3.23 (i). Hence, $\int \widetilde{f}_2 \widetilde{f}_2'\, d(\overline{\delta}_1 \otimes \delta_2) = J_{22}$ which proves part (i).

If $\mathbf{1} \notin \text{span}(a_1^{(1)}, ..., a_{p_1}^{(1)})$, then $\int a^{(1)}\, d\overline{\delta}_1 = \mathbf{0}$ and, hence, $\widetilde{L}_1 = \mathbf{0}$ by Lemma 3.23 (ii) which proves part (ii). □

In most situations which are considered here a constant term will be involved in the first marginal model as in most practical applications, such that the representation of Lemma 6.5 (i) applies. In particular, for model (6.6) exhibiting a one-way layout in the first factor we observe $\mathbf{1} = \mathbf{1}_I' a^{(1)} = \sum_{i=1}^I a_i^{(1)}$, hence $\mathbf{1} \in \text{span}(a_1^{(1)}, ..., a_I^{(1)})$. Furthermore

$\int a^{(1)}\, d\overline{\delta}_1 = \frac{1}{I}\mathbf{1}_I$, $\mathbf{I}_1(\overline{\delta}_1) = \frac{1}{I}\mathbf{E}_I$, and $\widetilde{L}_1 = \mathbf{1}'_I$. Thus for this class of models a generalized inverse is given by

$$\mathbf{I}(\overline{\delta}_1 \otimes \delta_2)^- = \begin{pmatrix} I\,\mathbf{E}_I \otimes (\int f_1 f_1'\, d\delta_2)^- + \mathbf{1}_I^I \otimes \widetilde{L}_2' J_{22}^- \widetilde{L}_2 & -\mathbf{1}_I \otimes \widetilde{L}_2' J_{22}^- \\ -\mathbf{1}'_I \otimes J_{22}^- \widetilde{L}_2 & J_{22}^- \end{pmatrix}.$$

The next result relates the identifiability in the two-factor model under a design δ with that of the corresponding symmetrized design $\overline{\delta} = \overline{\delta}_1 \otimes \delta_2$. To keep notations simple we restrict our attention to the components β_1 and β_2 in the two-factor model,

Lemma 6.6

Let T_1 be finite, $\overline{\delta}_1$ the uniform design on T_1, let G_1 induce linear transformations of $a^{(1)}$ let G_1 act transitively on T_1, and let $\psi(\beta) = \beta$ ($\psi(\beta) = \beta_1$, $\psi(\beta) = \beta_2$, respectively). If $\delta \in \Delta(\psi)$ then $\overline{\delta}_1 \otimes \delta_2 \in \Delta(\psi)$, where δ_2 is the second marginal of δ.

Proof

In view of Lemma 3.17 it remains to show that G induces linear transformations of the linear aspects under consideration. For the whole parameter vector β this is always true. For $\beta_1 = L_{\beta_1}\beta$, $L_{\beta_1} = (\mathbf{E}_{p_1} \otimes \mathbf{E}_{p_{2,1}}|\mathbf{0})$, these transformations are given by $L_{\beta_1}Q'_g = (Q'_{1,g_1} \otimes \mathbf{E}_{p_{2,1}}|\mathbf{0}) = (Q'_{1,g_1} \otimes \mathbf{E}_{p_{2,1}})L_{\beta_1}$ and, hence, $Q_{\beta_1,g} = (Q_{1,g_1} \otimes \mathbf{E}_{p_{2,1}})$, for every $g \in G$. Finally, β_2 is invariant with respect to G. □

We can now establish that the set of product designs $\overline{\delta}_1 \otimes \delta_2$, $\delta_2 \in \Delta_2$, constitute an essentially complete class, if we are interested only in the component β_2 of the parameter vector not depending on the first factor,

Theorem 6.7

Let T_1 be finite, $\overline{\delta}_1$ the uniform design on T_1, let G_1 induce linear transformations of $a^{(1)}$ and let G_1 act transitively on T_1. If $\delta \in \Delta(\beta_2)$ in the two-factor model $\mu(t_1, t_2) = (a^{(1)}(t_1) \otimes f_1(t_2))'\beta_1 + f_2(t_2)'\beta_2$, $(t_1, t_2) \in T_1 \times T_2$, then $\mathbf{C}_{\beta_2}(\delta) \geq \mathbf{C}_{\beta_2}(\overline{\delta}_1 \otimes \delta_2)$, where δ_2 is the marginal of δ with respect to the second factor.

Proof

By Lemma 6.6 we have $\overline{\delta}_1 \otimes \delta_2 \in \Delta(\beta_2)$. From Lemma 6.5 we obtain $\mathbf{C}_{\beta_2}(\overline{\delta}_1 \otimes \delta_2) = J_{22}^{-1} = \mathbf{C}_{2,\beta_2}(\delta_2)$ for the covariance matrix of the symmetrized design $\overline{\delta}_1 \otimes \delta_2$ in case $\mathbf{1} \in \mathrm{span}(a_1^{(1)}, ..., a_{p_1}^{(1)})$ and, alternatively, $\mathbf{C}_{\beta_2}(\overline{\delta}_1 \otimes \delta_2) = (\int f_2 f_2'\, d\delta_2)^{-1} = \mathbf{C}^{(2)}(\delta_2)$ if $\mathbf{1} \notin \mathrm{span}(a_1^{(1)}, ..., a_{p_1}^{(1)})$, where $\mathbf{C}^{(2)}(\delta_2)$ is the covariance matrix in the reduced second marginal model $\mu_2^{(2)}(t_2) = f_2'\beta_2$. For the first case we notice that the first marginal can be reparametrized in such a way that there is an explicit constant term. This causes a reparametrization of the whole model which does not affect β_2. Hence, the second marginal model can be regarded as a submodel of the whole two-factor model. Also, the reduced second marginal model is a submodel and the result follows in both cases by the refinement argument of Lemma 3.5. □

Optimum designs can now be characterized by their second marginal as follows

Corollary 6.8

Let T_1 be finite, $\overline{\delta}_1$ the uniform design on T_1, let G_1 induce linear transformations of $a^{(1)}$ and let G_1 act transitively on T_1.

(i) *If* $\mathbf{1} \in \text{span}(a_1^{(1)}, ..., a_{p_1}^{(1)})$ *and if* δ_2^* *is A- (D- resp. E-) optimum for* β_2 *in the marginal model* $\mu_2(t_2) = f_1(t_2)'\beta_1^{(2)} + f_2(t_2)'\beta_2$, $t_2 \in T_2$, *then the product design* $\delta^* = \overline{\delta}_1 \otimes \delta_2^*$ *is A- (D- resp. E-) optimum for* β_2 *in the two-factor model* $\mu(t_1, t_2) = (a^{(1)}(t_1) \otimes f_1(t_2))'\beta_1 + f_2(t_2)'\beta_2$, $(t_1, t_2) \in T_1 \times T_2$.

(ii) *If* $\mathbf{1} \notin \text{span}(a_1^{(1)}, ..., a_{p_1}^{(1)})$ *and if* δ_2^* *is A- (D- resp. E-) optimum for* β_2 *in the reduced marginal model* $\mu_2^{(2)}(t_2) = f_2(t_2)'\beta_2$, $t_2 \in T_2$, *then the product design* $\delta^* = \overline{\delta}_1 \otimes \delta_2^*$ *is A- (D- resp. E-) optimum for* β_2 *in the two-factor model* $\mu(t_1, t_2) = (a^{(1)}(t_1) \otimes f_1(t_2))'\beta_1 + f_2(t_2)'\beta_2$, $(t_1, t_2) \in T_1 \times T_2$.

Proof

In view of Theorem 6.7 it remains to show that $\overline{\delta}_1 \otimes \delta_2^* \in \Delta(\beta_2)$. As $\delta_2^* \in \Delta_2(\beta_2)$ there exists a matrix $M^{(2)} \in \mathbb{R}^{p_{2,2} \times p_2}$ with $(\mathbf{0}|\mathbf{E}_{p_{2,2}}) = M^{(2)}\mathbf{I}_2(\delta_2^*)$. We can partition $M^{(2)} = (M_1^{(2)}|M_2^{(2)})$ according to the components $\beta_1^{(2)}$ and β_2, i.e. $M_1^{(2)} \in \mathbb{R}^{p_{2,2} \times p_{2,1}}$, $M_2^{(2)} \in \mathbb{R}^{p_{2,2} \times p_{2,2}}$. If $\mathbf{1} \in \text{span}(a_1^{(1)}, ..., a_{p_1}^{(1)})$ there is a vector $\ell \in \mathbb{R}^{p_1}$ such that $\ell' a^{(1)}(t_1) = 1$, for all $t_1 \in T_1$. Let $M = (\ell' \otimes M_1^{(2)}|M_2^{(2)})$. Then $(\mathbf{0}|\mathbf{E}_{p_{2,2}}) = M\mathbf{I}(\overline{\delta}_1 \otimes \delta_2^*)$ and β_2 is identifiable under $\overline{\delta}_1 \otimes \delta_2^*$.

For $\mathbf{1} \notin \text{span}(a_1^{(1)}, ..., a_{p_1}^{(1)})$ the identifiability follows directly from the block diagonal structure of the information matrix $\mathbf{I}(\overline{\delta}_1 \otimes \delta_2^*)$ (see Lemma 6.5 (ii)). □

These results show that product designs are optimum for the component β_2 of the parameter vector belonging to the regression functions not depending on the levels t_1 of the first factor. Moreover, at least in the more important case of a constant term involved in the first marginal model, the second marginal of the optimum design is given by the corresponding optimum design in the marginal model. This last statement, however, does not remain true, in general, if we are interested in the components β_1 associated with the interacting regression functions or in the whole parameter vector β itself.

Theorem 6.9

Let T_1 *be finite,* $\overline{\delta}_1$ *the uniform design on* T_1, *let* G_1 *induce linear transformations of* $a^{(1)}$, *let* G_1 *act transitively on* T_1 *and let* $\mathbf{1} \in \text{span}(a_1^{(1)}, ..., a_{p_1}^{(1)})$.

(i) *If* δ_2^* *maximizes*

$$\det\left((\mathbf{I}_1(\overline{\delta}_1) - \textstyle\int a^{(1)}\, d\overline{\delta}_1(\int a^{(1)}\, d\overline{\delta}_1)') \otimes \int f_1 f_1'\, d\delta_2 + \int a^{(1)}\, d\overline{\delta}_1(\int a^{(1)}\, d\overline{\delta}_1)' \otimes \mathbf{C}_{2,\beta_1^{(2)}}(\delta_2)^{-1}\right)$$

in $\delta_2 \in \Delta_2(\beta_1^{(2)})$, *then the product design* $\delta^* = \overline{\delta}_1 \otimes \delta_2^*$ *is D-optimum for* β_1 *in the two-factor model* $\mu(t_1, t_2) = (a^{(1)}(t_1) \otimes f_1(t_2))'\beta_1 + f_2(t_2)'\beta_2$, $(t_1, t_2) \in T_1 \times T_2$.

(ii) *If* G_1 *is orthogonal for* $a^{(1)}$ *and if* δ_2^* *minimizes*

$$\gamma(\overline{\delta}_1)\text{tr}\left(\mathbf{C}_{2,\beta_1^{(2)}}(\delta_2)\right) + \left(\text{tr}(\mathbf{C}_1(\overline{\delta}_1)) - \gamma(\overline{\delta}_1)\right)\text{tr}\left((\textstyle\int f_1 f_1'\, d\delta_2)^{-1}\right)$$

in $\delta_2 \in \Delta_2(\beta_1^{(2)})$, *where* $\gamma(\overline{\delta}_1) = (\int a^{(1)}\, d\overline{\delta}_1)'\mathbf{C}_1(\overline{\delta}_1)^{-2} \int a^{(1)}\, d\overline{\delta}_1$, *then the product design* $\delta^* = \overline{\delta}_1 \otimes \delta_2^*$ *is A-optimum for* β_1 *in the two-factor model* $\mu(t_1, t_2) = (a^{(1)}(t_1) \otimes f_1(t_2))'\beta_1 + f_2(t_2)'\beta_2$, $(t_1, t_2) \in T_1 \times T_2$.

Proof

First we have to show that $\delta_2 \in \Delta_2(\beta_1^{(2)})$ implies $\overline{\delta}_1 \otimes \delta_2 \in \Delta(\beta_1)$. We transform the regression functions of the first marginal $a^{(1)}$ to $\widetilde{a}^{(1)}$ with $\widetilde{a}_1^{(1)} = \mathbf{1}$ and $\int \widetilde{a}_i^{(1)}\, d\overline{\delta}_1 = 0$, $i = 2, ..., p_1$. This induces a linear transformation of the parameters in the two-factor model, $\mu(t_1, t_2) = (\widetilde{a}^{(1)}(t_1) \otimes f_1(t_2))'\widetilde{\beta}_1 + f_2(t_2)'\beta_2$ which induces a linear transformations of β_1, i. e. $\widetilde{\beta}_1 = L_1\beta_1$ for some regular $L_1 \in \mathbb{R}^{p_1p_{2,1}\times p_1p_{2,1}}$. Hence β_1 is identifiable if and only if $\widetilde{\beta}_1$ is identifiable.

$$\widetilde{\mathbf{I}}(\overline{\delta}_1 \otimes \delta_2) = \begin{pmatrix} \int f_1 f_1'\, d\delta_2 & \mathbf{0} & \int f_1 f_2'\, d\delta_2 \\ \mathbf{0} & (\int \widetilde{a}_i^{(1)}\widetilde{a}_j^{(1)}\, d\overline{\delta}_1)_{i=2,...,p_1}^{j=2,...,p_1} \otimes \int f_2 f_2'\, d\delta_2 & \mathbf{0} \\ \int f_2 f_1'\, d\delta_2 & \mathbf{0} & \int f_2 f_2'\, d\delta_2 \end{pmatrix}$$

is the information matrix in the transformed model. Now, $\delta_2 \in \Delta_2(\beta_1^{(2)})$ implies that $\int f_1 f_1'\, d\delta_2$ is regular which together with the identifiability of $\beta_1^{(2)}$ under δ_2 yields $\overline{\delta}_1 \otimes \delta_2 \in \Delta(\widetilde{\beta}_1)$.

Next, we recall that G induces linear transformations of β_1 with transformation matrices $Q_{\beta_1,g} = Q_{1,g_1} \otimes \mathbf{E}_{p_{2,1}}$ for every $g = (g_1, \mathrm{id}|_{T_2}) \in G$ (see the proof of Lemma 6.6).

(i) The D_{β_1}-criterion is invariant by Theorem 3.18. Because of the quasi-convexity of the D_{β_1}-criterion we have to look for the best product design with its first marginal $\overline{\delta}_1$ uniform on T_1.

If we center $a^{(1)} \otimes f_1$ on f_2, i. e.

$$\widetilde{a}(t_1, t_2) = \begin{pmatrix} a^{(1)}(t_1) \otimes f_1(t_2) - \int a^{(1)}\, d\overline{\delta}_1 \otimes \int f_1 f_2'\, d\delta_2 (\int f_2 f_2'\, d\delta_2)^- f_2(t_2) \\ f_2(t_2) \end{pmatrix}$$

instead of the reverse operation used in the proof of Lemma 6.5 (i) we obtain the transformed information matrix

$$\widetilde{\mathbf{I}}(\overline{\delta}_1 \otimes \delta_2) = \begin{pmatrix} J_1 & \mathbf{0} \\ \mathbf{0} & \int f_2 f_2'\, d\delta_2 \end{pmatrix}$$

where $J_1 = \mathbf{I}_1(\overline{\delta}_1) \otimes \int f_1 f_1'\, d\delta_2 - \int a^{(1)}\, d\overline{\delta}_1 (\int a^{(1)}\, d\overline{\delta}_1)' \otimes \int f_1 f_2'\, d\delta_2 (\int f_2 f_2'\, d\delta_2)^- \int f_2 f_1'\, d\delta_2$. Hence, $\mathbf{C}_{\beta_1}(\overline{\delta}_1 \otimes \delta_2) = J_1^{-1}$, $\mathbf{C}_{2,\beta_1^{(2)}}(\delta_2)^{-1} = \int f_1 f_1'\, d\delta_2 - \int f_1 f_2'\, d\delta_2 (\int f_2 f_2'\, d\delta_2)^- \int f_2 f_1'\, d\delta_2$ and the inverse of the covariance matrix $\mathbf{C}_{\beta_1}(\overline{\delta}_1 \otimes \delta_2)$ attains the form presented in the theorem which proves the D-optimality of the product design $\overline{\delta}_1 \otimes \delta_2$.

(ii) From the orthogonality of G_1 it follows that $Q_{\beta_1,g}$ is orthogonal for every $g \in G$. Hence, the A_{β_1}-criterion is invariant with respect to G by Theorem 3.18. Thus, we can restrict our interest to the essentially complete class of invariant designs because of the convexity of the A_{β_1}-criterion (see Theorem 3.13).

The invariant designs are given by $\overline{\delta}_1 \otimes \delta_2$, $\delta_2 \in \Delta_2$, according to Lemma 6.4. By the representation established in Lemma 6.5 (i) and with the notations used there we obtain $\mathbf{C}_{\beta_1}(\overline{\delta}_1 \otimes \delta_2) = \mathbf{C}_1(\overline{\delta}_1) \otimes (\int f_1 f_1'\, d\delta_2)^{-1} + \widetilde{L}_1'\widetilde{L}_1 \otimes \widetilde{L}_2' J_{22}^- \widetilde{L}_2$ for every $\overline{\delta}_1 \otimes \delta_2 \in \Delta(\beta_1)$. As $\mathbf{C}_{2,\beta_1^{(2)}}(\delta_2) = (\int f_1 f_1'\, d\delta_2)^{-1} - \widetilde{L}_2' J_{22}^- \widetilde{L}_2$ and $\mathrm{tr}(\widetilde{L}_1'\widetilde{L}_1) = \gamma(\overline{\delta}_1)$ this yields $\mathrm{tr}(\mathbf{C}_{\beta_1}(\overline{\delta}_1 \otimes \delta_2)) = \gamma(\overline{\delta}_1)\mathrm{tr}(\mathbf{C}_{2,\beta_1^{(2)}}(\delta_2)) + (\mathrm{tr}(\mathbf{C}_1(\overline{\delta}_1)) - \gamma(\overline{\delta}_1))\mathrm{tr}(\int f_1 f_1'\, d\delta_2)^{-1}$ and the product design $\overline{\delta}_1 \otimes \delta_2^*$ is A_{β_1}-optimum. □

Although Theorem 6.9 offers an explicit characterization of an optimum design it is still difficult to solve the extremum problems for determining δ_2^*, in general. A simplification is presented in Corollary 6.14 in case the underlying model is given by (6.6) where the first marginal model is a one-way layout. For the sake of completeness we also mention some results for the case that no constant term is involved in $a^{(1)}$. As the proof is analogous to the previous one it will be omitted,

Theorem 6.10

Let T_1 be finite, $\overline{\delta}_1$ the uniform design on T_1, let G_1 induce linear transformations of $a^{(1)}$, let G_1 act transitively on T_1 and let $\mathbf{1} \notin \text{span}(a_1^{(1)}, ..., a_{p_1}^{(1)})$.

(i) *If δ_2^* is D-optimum in the reduced marginal model $\mu_2^{(1)}(t_2) = f_1(t_2)'\beta^{(2)}$, $t_2 \in T_2$, then the product design $\delta^* = \overline{\delta}_1 \otimes \delta_2^*$ is D-optimum for β_1 in the two-factor model $\mu(t_1, t_2) = (a^{(1)}(t_1) \otimes f_1(t_2))'\beta_1 + f_2(t_2)'\beta_2$, $(t_1, t_2) \in T_1 \times T_2$.*

(ii) *If G_1 is orthogonal for $a^{(1)}$ and if δ_2^* is A-optimum in the reduced marginal model $\mu_2^{(1)}(t_2) = f_1(t_2)'\beta^{(2)}$, $t_2 \in T_2$, then the product design $\delta^* = \overline{\delta}_1 \otimes \delta_2^*$ is A-optimum for β_1 in the two-factor model $\mu(t_1, t_2) = (a^{(1)}(t_1) \otimes f_1(t_2))'\beta_1 + f_2(t_2)'\beta_2$, $(t_1, t_2) \in T_1 \times T_2$.*

We will now investigate the whole parameter vector β. The equivalence of the identifiability of β under $\overline{\delta}_1 \otimes \delta_2$ and of $\beta^{(2)}$ under δ_2 can be derived by a combination of the results presented in the proofs of Corollary 6.8 and Theorem 6.9. In contrast to that we present a more straightforward proof, next,

Lemma 6.11

Let T_1 be finite, $\overline{\delta}_1$ the uniform design on T_1, let G_1 induce linear transformations of $a^{(1)}$ and let G_1 act transitively on T_1.

(i) *If $\mathbf{1} \in \text{span}(a_1^{(1)}, ..., a_{p_1}^{(1)})$, then $\overline{\delta}_1 \otimes \delta_2 \in \Delta(\beta)$ if and only if $\delta_2 \in \Delta_2(\beta^{(2)})$.*

(ii) *If $\mathbf{1} \notin \text{span}(a_1^{(1)}, ..., a_{p_1}^{(1)})$, then $\overline{\delta}_1 \otimes \delta_2 \in \Delta(\beta)$ if and only if $\int f_1 f_1' \, d\delta_2$ and $\int f_2 f_2' \, d\delta_2$ are regular.*

Proof

(i) As the second marginal model can be regarded as a submodel of the two-factor model (see Theorem 6.7) the identifiability of β implies that of $\beta^{(2)}$ under the marginal design δ_2. Conversely, if $\delta_2 \in \Delta_2(\beta^{(2)})$ then $\mathbf{I}_1(\overline{\delta}_1) \otimes \mathbf{I}_2(\delta_2) > \mathbf{0}$ by Lemma 4.1. Let

$$L_0 = \begin{pmatrix} \mathbf{E}_{p_1} \otimes \mathbf{E}_{p_{2,1}} & \mathbf{0} \\ \mathbf{0} & \ell' \otimes \mathbf{E}_{p_{2,2}} \end{pmatrix},$$

then we obtain $L_0(\mathbf{I}_1(\overline{\delta}_1) \otimes \mathbf{I}_2(\delta_2))L_0' = \mathbf{I}(\overline{\delta}_1 \otimes \delta_2)$ and, hence, $\mathbf{I}(\overline{\delta}_1 \otimes \delta_2) > \mathbf{0}$ which proves the identifiability of β.

(ii) This follows directly from the block diagonal structure of the information matrix (Lemma 6.5). □

We are now able to characterize optimum designs for the whole parameter vector,

Theorem 6.12

Let T_1 be finite, $\overline{\delta}_1$ the uniform design on T_1, let G_1 induce linear transformations of $a^{(1)}$, let G_1 act transitively on T_1 and let $\mathbf{1} \in \text{span}(a_1^{(1)}, ..., a_{p_1}^{(1)})$.

(i) *If δ_2^* maximizes*

$$\det\left(\int f_1 f_1' \, d\delta_2\right)^{p_1 - 1} \det\left(\mathbf{I}_2(\delta_2)\right)$$

in $\delta_2 \in \Delta_2(\beta^{(2)})$, *then the product design* $\delta^* = \overline{\delta}_1 \otimes \delta_2^*$ *is D-optimum in the two-factor model* $\mu(t_1,t_2) = (a^{(1)}(t_1) \otimes f_1(t_2))'\beta_1 + f_2(t_2)'\beta_2$, $(t_1,t_2) \in T_1 \times T_2$.

(ii) *If* G_1 *is orthogonal for* $a^{(1)}$ *and if* δ_2^* *minimizes*

$$(\mathrm{tr}(\mathbf{C}_1(\overline{\delta}_1)) - \gamma(\overline{\delta}_1))\,\mathrm{tr}\,((\textstyle\int f_1 f_1'\,d\delta_2)^{-1}) + \gamma(\overline{\delta}_1)\mathrm{tr}\,(\mathbf{C}_{2,\psi_A}(\delta_2))\ ,$$

where $\gamma(\overline{\delta}_1) = (\int a^{(1)}\,d\overline{\delta}_1)'\mathbf{C}_1(\overline{\delta}_1)^{-2}\int a^{(1)}\,d\overline{\delta}_1$ *and* $\psi_A(\beta^{(2)}) = (\beta_1^{(2)'}, \gamma(\overline{\delta}_1)^{-1/2}\beta_2')'$, *in* $\delta_2 \in \Delta_2(\beta^{(2)})$, *then the product design* $\delta^* = \overline{\delta}_1 \otimes \delta_2^*$ *is A-optimum in the two-factor model* $\mu(t_1,t_2) = (a^{(1)}(t_1) \otimes f_1(t_2))'\beta_1 + f_2(t_2)'\beta_2$, $(t_1,t_2) \in T_1 \times T_2$.

Proof

(i) The D-criterion is invariant and, hence, by the quasi-convexity of this criterion an optimum design can be found in the essentially complete class of invariant designs (Theorem 3.13) which are of the form $\overline{\delta}_1 \otimes \delta_2$, $\delta_2 \in \Delta_2$ (Lemma 6.4).

Using the notation $\widetilde{\mathbf{I}}(\overline{\delta}_1 \otimes \delta_2)$ for the transformed information matrix for f_2 centered on f_1 as introduced in the proof of Lemma 6.5 (i) we find

$$\det(\mathbf{I}(\overline{\delta}_1 \otimes \delta_2)) = \det(\widetilde{\mathbf{I}}(\overline{\delta}_1 \otimes \delta_2)) = \det(\mathbf{I}_1(\overline{\delta}_1))^{p_{2,1}} \det(\textstyle\int f_1 f_1'\,d\delta_2)^{p_1} \det(\mathbf{C}_{2,\beta_2}(\delta_2))^{-1}\ .$$

As $\det(\mathbf{I}_2(\delta_2)) = \det(\int f_1 f_1'\,d\delta_2)\det(\mathbf{C}_{2,\beta_2}(\delta_2))^{-1}$ the product design $\overline{\delta}_1 \otimes \delta_2^*$ is D-optimum.

(ii) By the orthogonality of G_1 all Q_g are orthogonal, $g \in G$, and the A-criterion is invariant. Hence, the best symmetrized product design $\overline{\delta}_1 \otimes \delta_2$ will be A-optimum.

By Lemma 6.5 (i) and with the notation introduced there we obtain

$$\mathrm{tr}(\mathbf{C}(\overline{\delta}_1 \otimes \delta_2)) = \mathrm{tr}(\mathbf{C}_1(\overline{\delta}_1))\,\mathrm{tr}((\textstyle\int f_1 f_1'\,d\delta_2)^{-1}) + \widetilde{L}_1\widetilde{L}_1'\mathrm{tr}(\widetilde{L}_2'\mathbf{C}_{2,\beta_2}(\delta_2)\widetilde{L}_2) + \mathrm{tr}(\mathbf{C}_{2,\beta_2}(\delta_2))\ .$$

for every product design $\overline{\delta}_1 \otimes \delta_2 \in \Delta(\beta)$. Because of $\widetilde{L}_1\widetilde{L}_1' = \gamma(\overline{\delta}_1)$ and

$$\mathbf{C}_{2,\psi_A}(\delta_2) = \begin{pmatrix} (\int f_1 f_1'\,d\delta_2)^{-1} + \widetilde{L}_2'\mathbf{C}_{2,\beta_2}(\delta_2)\widetilde{L}_2 & -\gamma(\overline{\delta}_1)^{-1/2}\widetilde{L}_1' \otimes \widetilde{L}_2'\mathbf{C}_{2,\beta_2}(\delta_2) \\ -\gamma(\overline{\delta}_1)^{-1/2}\widetilde{L}_1 \otimes \mathbf{C}_{2,\beta_2}(\delta_2)\widetilde{L}_2 & \gamma(\overline{\delta}_1)^{-1}\mathbf{C}_{2,\beta_2}(\delta_2) \end{pmatrix}$$

the product design $\overline{\delta}_1 \otimes \delta_2^*$ is A-optimum. □

Note that for the D-optimality the optimization problem which occurs in Theorem 6.12 (i) can be rewritten in terms of a criterion introduced by Läuter (1974) for mixtures of models. In the particular case of a polynomial regression in the second marginal model solutions to that optimization problem have been obtained by Dette (1990).

For the alternative case that no constant term is present in the first marginal $a^{(1)}$ the proof will be omitted, again, because it can be performed along the same lines as before.

Theorem 6.13

Let T_1 *be finite,* $\overline{\delta}_1$ *the uniform design on* T_1, *let* G_1 *induce linear transformations of* $a^{(1)}$, *let* G_1 *act transitively on* T_1 *and let* $\mathbf{1} \notin \mathrm{span}(a_1^{(1)},...,a_{p_1}^{(1)})$.

(i) *If* δ_2^* *maximizes*

$$\det\,(\textstyle\int f_1 f_1'\,d\delta_2)^{p_1}\,\det(\int f_2 f_2'\,d\delta_2)$$

in $\delta_2 \in \Delta_2(\beta^{(2)})$, *then the product design* $\delta^* = \overline{\delta}_1 \otimes \delta_2^*$ *is D-optimum in the two-factor model* $\mu(t_1,t_2) = (a^{(1)}(t_1) \otimes f_1(t_2))'\beta_1 + f_2(t_2)'\beta_2$, $(t_1,t_2) \in T_1 \times T_2$.

(ii) *If G_1 is orthogonal for $a^{(1)}$ and if δ_2^* minimizes*

$$\mathrm{tr}\left(\mathbf{C}_1(\overline{\delta}_1)\right)\mathrm{tr}\left(\left(\int f_1 f_1' \, d\delta_2\right)^{-1}\right) + \mathrm{tr}\left(\left(\int f_2 f_2' \, d\delta_2\right)^{-1}\right)$$

in $\delta_2 \in \Delta_2(\beta^{(2)})$, then the product design $\delta^* = \overline{\delta}_1 \otimes \delta_2^*$ is A-optimum in the two-factor model $\mu(t_1, t_2) = (a^{(1)}(t_1) \otimes f_1(t_2))'\beta_1 + f_2(t_2)'\beta_2$, $(t_1, t_2) \in T_1 \times T_2$.

Applications of this result are indicated in Examples 6.9 and 6.12.

In case of an underlying model (6.6) where the first factor is modeled by a one-way layout the conditions on the optimum marginal designs can be substantially simplified,

Corollary 6.14

Consider the two-factor model (6.6), $\mu(i,u) = f_1(u)'\beta_{1,i} + f_2(u)'\beta_2$, $i \in \{1,...,I\}$, $u \in U$, and let $\overline{\delta}_1$ be the uniform design on $\{1,...,I\}$. Then the product design $\overline{\delta}_1 \otimes \delta_2^*$ is

(i) *D-optimum for β if δ_2^* maximizes* $\det(\mathbf{I}_2(\delta_2))\det(\int f_1 f_1' \, d\delta_2)^{I-1}$

(ii) *D-optimum for $\beta_1 = (\beta_{1,i})_{i=1,...,I}$ if δ_2^* maximizes*

$$\det\left(\left(I\mathbf{E}_I - \mathbf{1}_I^I\right) \otimes \int f_1 f_1' \, d\delta_2 + \mathbf{1}_I^I \otimes \mathbf{C}_{2,\beta_1^{(2)}}(\delta_2)^{-1}\right)$$

(iii) *D-optimum for β_2 if δ_2^* is D-optimum for β_2*

(iv) *A-optimum for β if δ_2^* minimizes* $(I-1)\mathrm{tr}\left(\left(\int f_1 f_1' \, d\delta_2\right)^{-1}\right) + \mathrm{tr}\left(\mathbf{C}_{2,\psi_A}(\delta_2)\right)$, *where* $\psi_A(\beta^{(2)}) = (\beta_1^{(2)\prime}, \frac{1}{\sqrt{I}}\beta_2')'$,

(v) *A-optimum for $\beta_1 = (\beta_{1,i})_{i=1,...,I}$ if δ_2^* minimizes*

$$(I-1)\mathrm{tr}\left(\left(\int f_1 f_1' \, d\delta_2\right)^{-1}\right) + \mathrm{tr}\left(\mathbf{C}_{2,\beta_1^{(2)}}(\delta_2)\right)$$

(vi) *A-optimum for β_2 if δ_2^* is A-optimum for β_2*

in the single factor model $\mu_2(u) = f_1(u)'\beta_1^{(2)} + f_2(u)'\beta_2$, $u \in U$.

Proof

For the group G_1 acting on $\{1,...,I\}$ we may take the group of permutations which is orthogonal for $a^{(1)}$ and the uniform design $\overline{\delta}_1$ is invariant with respect to G_1.

In the marginal model of a one-way layout we have $\int a^{(1)} \, d\overline{\delta}_1 = \frac{1}{I}\mathbf{1}_I$, $\mathbf{I}_1(\overline{\delta}_1) = \frac{1}{I}\mathbf{E}_I$ and, hence, $\gamma(\overline{\delta}_1) = I$. Inserting these expressions in the corresponding formulae we obtain (i) to (vi) from Theorem 6.12, Theorem 6.9 and Corollary 6.8, respectively. □

Example 6.5 *One-way layout combined with linear regression: Common intercept:*

We consider the two-factor model $\mu(i,u) = \beta_0 + \beta_{1,i}u$ (see Example 6.4 c)). The appropriately partitioned regression functions for the second factor are given by $f_1(u) = u$ and $f_2 = \mathbf{1}$.

(i) First, we investigate the case of a standardized marginal design region $U = [-1, 1]$. In the second marginal model the design δ_2^* which assigns equal weights $\frac{1}{2}$ to both end-points -1 and 1 of the interval is D- and A-optimum (see Example 3.9). The same design also maximizes $\int f_1^2 \, d\delta_2$ and minimizes simultaneously $\mathbf{C}_{2,\beta_1^{(2)}}(\delta_2)$ and $\mathrm{tr}(\mathbf{C}_{2,\psi_A}(\delta_2))$. The latter three properties can be seen by $\int f_1^2 \, d\delta_2^* = 1 \geq \int f_1^2 \, d\delta_2$, $\mathbf{C}_{2,\beta_1^{(2)}}(\delta_2^*) = 1 \leq \mathbf{C}_{2,\beta_1^{(2)}}(\delta_2)$ and $\mathrm{tr}(\mathbf{C}_{2,\psi_A}(\delta_2^*)) = 1 + \frac{1}{I} \leq \mathrm{tr}(\mathbf{C}_{2,\psi_A}(\delta_2))$, respectively. Moreover, δ_2^* is also optimum for the constant term.

In view of Corollary 6.14 this establishes the simultaneous D-, A-, and A_{β_1}-optimality of the product design $\overline{\delta}_1 \otimes \delta_2^*$ and its optimality for the constant term β_0.

As $\mathbf{C}_{2,\beta_1^{(2)}}(\delta_2^*)^{-1} - \int f_1^2\, d\delta_2^* = 0 \geq \mathbf{C}_{2,\beta_1^{(2)}}(\delta_2)^{-1} - \int f_1^2\, d\delta_2$ we obtain $\int f_1^2\, d\delta_2^*(I\mathbf{E}_I - \mathbf{1}_I^I) + \mathbf{C}_{2,\beta_1^{(2)}}(\delta_2^*)^{-1}\mathbf{1}_I^I = I\mathbf{E}_I \geq \int f_1^2\, d\delta_2(I\mathbf{E}_I - \mathbf{1}_I^I) + \mathbf{C}_{2,\beta_1^{(2)}}(\delta_2)^{-1}\mathbf{1}_I^I$ and, hence, the D_{β_1}-optimality of the same design $\overline{\delta}_1 \otimes \delta_2^*$.

(ii) If the underlying marginal design region is symmetric, i. e. $U = [-u_0, u_0]$, the scale transformation does not affect the arguments used in (i) and an optimum design with respect to all criteria under consideration is again given by $\overline{\delta}_1 \otimes \delta_2^*$ where δ_2^* is the optimum marginal design which assigns equal weights $\frac{1}{2}$ to each endpoint $-u_0$ and u_0.

(iii) For asymmetric design regions $U = [u_1, u_2]$ this result does not carry over even for the D-criterion, because the model is not linearly transformed by translations. For example for $U = [0, u_2]$ a D-optimum design is given by $\overline{\delta}_1 \otimes \delta_2^*$ where δ_2^* assigns weight $\frac{I}{I+1}$ to u_2 and weight $\frac{1}{I+1}$ to 0. In particular, by a majorization argument which is explained in Example 3.9 we see that a marginal design δ_2^* generating a D-optimum product design $\overline{\delta}_1 \otimes \delta_2^*$ is concentrated on the endpoints for any interval U. The optimum weights depend only on the relation $\gamma = |u_2/u_1|$, e. g. the optimum weight at the right endpoint u_2 can be calculated as $2/(2 + \sqrt{I^2(\gamma^2-1)^2 + 4\gamma^2} - I(\gamma^2 - 1))$. □

If the factor described by the one-way layout only affects the constant term of the second marginal model, i. e. the factors are non-interacting as treated in Section 5, the characterization of optimum designs becomes much easier to survey (see WIERICH (1986b) and WIERICH (1988a)):

Corollary 6.15

Consider the two-factor model $\mu(i, u) = \alpha_i + f_2(u)'\beta_2$, $i \in \{1, ..., I\}$, $u \in U$, and let $\overline{\delta}_1$ be the uniform design on $\{1, ..., I\}$. Then the product design $\overline{\delta}_1 \otimes \delta_2^*$ is

(i) *D-optimum if δ_2^* is D-optimum*

(ii) *D_α-optimum for $\alpha = (\alpha_i)_{i=1,...,I}$ if δ_2^* is β_0-optimum*

(iii) *D_{β_2}-optimum if δ_2^* is D_{β_2}-optimum*

(iv) *A-optimum if δ_2^* is A_ψ-optimum where $\psi(\beta^{(2)}) = (\sqrt{I}\beta_0, \beta_2')'$*

(v) *A_α-optimum for $\alpha = (\alpha_i)_{i=1,...,I}$ if δ_2^* is β_0-optimum*

(vi) *A_{β_2}-optimum if δ_2^* is A_{β_2}-optimum*

in the single factor model $\mu_2(u) = \beta_0 + f_2(u)'\beta_2$, $u \in U$.

Proof

Because of $f_1 = \mathbf{1}$ the characterizations follow directly from Corollary 6.14. □

For the present parametrization note that there are A-optimum product designs in contrast to the results of Section 5. The characterizations of D-optimum designs can also be obtained from Theorem 5.2 for β, from Corollary 5.11 for β_2 and from Corollary 5.18 for α, respectively, as those criteria are not affected by the reparametrization.

Example 6.6 *Two-way layout without interactions (cf Example 3.7):*

We consider a standard-control parametrization for the model $\mu(i, j) = \alpha_i^{(1)} + \alpha_j^{(2)}$, $i = 1, ..., I$, $j = 1, ..., J$, of two qualitative factors without interactions, with identifiability condition $\alpha_J^{(2)} = 0$. As shown in Example 5.4 the uniform design $\overline{\delta}_1 \otimes \overline{\delta}_2$ is D-optimum which assigns equal weights $\frac{1}{IJ}$ to each level combination (i, j). By means of Corollary 6.15 we are now able to determine the A-optimum design $\overline{\delta}_1 \otimes \delta_2^*$ for which the marginal design δ_2^* assigns weight $\frac{1}{J-1+\sqrt{I+J-1}}$ to the treatment levels $j = 1, ..., J-1$ and weight $\frac{\sqrt{I+J-1}}{J-1+\sqrt{I+J-1}}$

to the control level J. We want to add that in view of Example 5.7 the A-optimality of product designs cannot be expexted, in general, for standard-control parametrization in higher dimensions. For the optimality and non-optimality of product designs in the K-way layout without interactions in standard-control parametrization in case $K > 2$ we refer to WIERICH (1988b). □

Example 6.7 *One-way layout combined with linear regression: Common slope:*

In this model $\mu(i,u) = \alpha_i + \beta_2 u$, $i = 1,...,I$, $u \in U$ (see Example 6.4 b)), only the constant term depends on the first factor. If the marginal design region U is symmetric, i. e. $U = [-u_0, u_0]$, the product design $\overline{\delta}_1 \otimes \delta_2^*$ in which δ_2^* assigns equal weights $\frac{1}{2}$ to each of the endpoints $-u_0$ and u_0 of the interval is A- and D-optimum according to Example 5.14. In view of the characterizations given in Corollary 6.15 the same design is also D_α-, D_{β_2}-, A_α-, and A_{β_2}-optimum.

We note that for arbitrary intervals U the design $\overline{\delta}_1 \otimes \delta_2^*$ where δ_2^* assigns equal weight $\frac{1}{2}$ to each endpoint remains D and D_{β_2}-optimum because both the model and the linear aspect β_2 is linearly transformed by translations. □

If the second marginal model is more complicated it is unlikely to find a design which is simultaneously optimum for different underlying interaction structures in contrast to the situations treated in Examples 6.4, 6.5 and 6.7,

Example 6.8 *One-way layout combined with quadratic regression:*

If we have a quadratic regression $\mu_2(u) = \beta_0 + \beta_1 u + \beta_2 u^2$ in the second marginal model, then there are a couple of possible different interaction structures with a one-way layout describing the first factor (see Table 6.1, cf KUROTSCHKA (1988) and KUROTSCHKA, SCHWABE, and WIERICH (1992))

Let us assume that the design region U is standardized for the second factor, $U = [-1, 1]$. What we know so far is that for any of these interaction structures there is a symmetrized D- (or A-) optimum design $\overline{\delta}_1 \otimes \delta_2$.

The criteria given in Corollaries 6.14 and 6.15 are weighted sums of criteria based on the information matrices (resp. covariance matrices) of the full marginal model and its reduced submodels. Those partial criteria are invariant with respect to sign change and all those criteria decrease as the information matrix increases. Hence, as in Example 3.9, every design is dominated by a symmetric design $\{\delta_2^{(w_2)}; 0 \leq w_2 \leq \frac{1}{2}\}$ supported on $\{-1, 0, 1\}$, where w_2 is the weight assigned to each of the endpoints -1 and 1 of the interval and $1 - 2w_2$ is the weight at the midpoint 0, and the criteria under consideration share this property. Therefore the whole design problem is reduced to finding an optimum weight w_2^* which can be solved explicitly by standard methods.

In Table 6.1 the optimum weights $1 - 2w_2^*$ assigned to 0 are listed for the optimum designs $\delta^* = \overline{\delta}_1 \otimes \delta_2^*$ with $\delta_2^* = \delta_2^{(w_2^*)}$. The first entry corresponds to pure quadratic regression, the second and third entries are the cases of complete and no interaction and the D-optimum weights $\frac{1}{3}$ there can be obtained by the results of Sections 4 and 5. However, more complicated interaction structures result in different optimum weights depending on the number I of levels of the qualitative factor as can be seen from the table. In particular, if I is large these differences are substantial.

Because all the models considered are linearly transformed by scale transformations

$\mu(i,u)$	D-optimum	A-optimum
1 $\beta_0 + \beta_1 u + \beta_2 u^2$	$\frac{1}{3}$	$\frac{1}{2}$
2 $\beta_{0i} + \beta_{1i} u + \beta_{2i} u^2$	$\frac{1}{3}$	$\frac{1}{2}$
3 $\beta_{0i} + \beta_1 u + \beta_2 u^2$	$\frac{1}{3}$	$\frac{\sqrt{I+1}}{\sqrt{I+1}+\sqrt{2}}$
4 $\beta_0 + \beta_{1i} u + \beta_2 u^2$	$\frac{1}{I+2}$	$\frac{\sqrt{2}}{\sqrt{2}+\sqrt{I^2+1}}$
5 $\beta_0 + \beta_1 u + \beta_{2i} u^2$	$\frac{1}{I+2}$	$\frac{\sqrt{I+1}}{\sqrt{I+1}+\sqrt{I^2+1}}$
6 $\beta_{0i} + \beta_{1i} u + \beta_2 u^2$	$\frac{1}{I+2}$	$\frac{\sqrt{I+1}}{\sqrt{I+1}+\sqrt{I^2+1}}$
7 $\beta_0 + \beta_{1i} u + \beta_{2i} u^2$	$\frac{1}{2I+1}$	$\frac{\sqrt{I+1}}{\sqrt{I+1}+I\sqrt{2}}$
8 $\beta_{0i} + \beta_1 u + \beta_{2i} u^2$	$\frac{I}{2I+1}$	$\frac{I\sqrt{2}}{I\sqrt{2}+\sqrt{I^2+1}}$

Table 6.1: One-way layout combined with quadratic regression, optimal weights at 0

the transformed design $\delta^* = \overline{\delta}_1 \otimes \delta_2^{(w_2^*)}$ where again $\delta_2^{(w_2^*)}$ assigns weight w_2^* to each of the endpoints $-u_0$ and u_0 and weight $1 - 2w_2^*$ to the midpoint 0 is seen to be D-optimum if the underlying design region is given by $U = [-u_0, u_0]$. For an assymetric design region U this transformation will produce a D-optimum design only in case that the model is linearly transformed by translations (besides for the entries 1 to 3 this is also true for entry 6 in Table 6.1). □

In contrast to the representations of the determinant and the trace of the covariance matrices it is much harder to express the eigenvalues in terms of the marginal model (for an exception see Example 6.12). Hence there are no results on E-optimality, yet. In general, there are also negative results concerning Q-optimality even with respect to a product weighting measure $\xi = \overline{\delta}_1 \otimes \xi_2$ which is invariant.

If we go back to Table 6.1 we see that the optimum weights in the second marginal tend to those corresponding to an optimum design for the reduced marginal model $\mu_2^{(1)}(u) = f_1(u)'\beta_1^{(2)}$, $u \in U$, in which only those regression functions are involved which interact with the first factor, as the number of levels I increases. This observation can be extended as follows,

Theorem 6.16

Let $(a^{(1,n)})_{n\in\mathbb{N}}$ be a sequence of regression functions $a^{(1,n)} : T_{1,n} \to \mathbb{R}^{p_{1,n}}$, $T_{1,n}$ finite, let a group $G_{1,n}$ induce linear transformations of $a^{(1,n)}$, let $G_{1,n}$ act transitively on $T_{1,n}$, let

$\overline{\delta}_{1,n}$ be the uniform design on $T_{1,n}$, let $\mathbf{C}_1^{(n)}(\overline{\delta}_{1,n}) = (\int a^{(1,n)}a^{(1,n)\prime}\, d\overline{\delta}_{1,n})^{-1}$ and let $\gamma_n(\overline{\delta}_{1,n}) = (\int a^{(1,n)}\, d\overline{\delta}_{1,n})'\mathbf{C}_1^{(n)}(\overline{\delta}_{1,n})^2 \int a^{(1,n)}\, d\overline{\delta}_{1,n}$, for every $n \in \mathbb{N}$.

If $(\delta^*_{2,n})_{n\in\mathbb{N}}$ is a sequence of designs on T_2, such that for every $n \in \mathbb{N}$ the design $\overline{\delta}_{1,n} \otimes \delta^*_{2,n}$ is D- (resp. A-) optimum in the nth model

$$\mu^{(n)}(t_1, t_2) = (a^{(1,n)}(t_1) \otimes f_1(t_2))'\beta_{n,1} + f_2(t_2)'\beta_2 ,$$

$(t_1, t_2) \in T_{1,n} \times T_2$ (with common second marginal model $\mu_2(t_2) = f_1(t_2)'\beta_1^{(2)} + f_2(t_2)'\beta_2$ and reduced second marginal model $\mu_2^{(1)}(t_2) = f_1(t_2)'\beta_1^{(2)}$, $t_2 \in T_2$), then the following implications hold

(i) *If $\overline{\delta}_{1,n} \otimes \delta^*_{2,n}$ is D-optimum for every n, if $\delta^*_{2,n} \xrightarrow{\mathcal{D}} \widetilde{\delta}_2$, i. e. if $\delta^*_{2,n}$ converges weakly to $\widetilde{\delta}_2$, and if $p_{1,n} \to \infty$, then $\widetilde{\delta}_2$ is D-optimum in the reduced marginal model.*

(ii) *If $\overline{\delta}_{1,n} \otimes \delta^*_{2,n}$ is A-optimum for every n, if $\delta^*_{2,n} \xrightarrow{\mathcal{D}} \widetilde{\delta}_2$, if $\mathrm{tr}(\mathbf{C}_1^{(n)}(\overline{\delta}_{1,n})) \to \infty$, and if $\mathrm{tr}(\mathbf{C}_1^{(n)}(\overline{\delta}_{1,n}))^{-1}\gamma_n(\overline{\delta}_{1,n}) \to 0$, then $\widetilde{\delta}_2$ is A-optimum in the reduced marginal model.*

(iii) *Conversely, if there is a unique D- (resp. A-) optimum design δ_2^* in the reduced marginal model and if $p_{1,n} \to \infty$ ($\mathrm{tr}(\mathbf{C}_1^{(n)}(\overline{\delta}_{1,n})) \to \infty$ and $\mathrm{tr}(\mathbf{C}_1^{(n)}(\overline{\delta}_{1,n}))^{-1}\gamma_n(\overline{\delta}_{1,n}) \to 0$, respectively), then $\delta^*_{2,n} \xrightarrow{\mathcal{D}} \delta_2^*$.*

Proof

(i) Denote by $\mathbf{I}^{(n)}$ and $\mathbf{I}_1^{(n)}$ the information matrices in the nth model and its corresponding first marginal model $\mu_1^{(n)}(t_1) = a^{(1,n)}(t_1)'\beta^{(1)}$, $t_1 \in T_{1,n}$.

First we assume that $\delta_2 \in \Delta_2(\beta^{(2)})$ and, additionally, $\mathbf{1} \in \mathrm{span}(a_1^{(1,n)}, ..., a_{p_{1,n}}^{(1,n)})$ for every $n \in \mathbb{N}$. As in the proof of Theorem 6.12 (ii) we obtain from Lemma 6.5 (i) that

$$\det(\mathbf{I}^{(n)}(\overline{\delta}_1 \otimes \delta_2)) = \det(\mathbf{I}_1^{(n)}(\overline{\delta}_1))^{p_{2,1}} \det(\textstyle\int f_1 f_1'\, d\delta_2)^{p_{1,n}-1} \det(\mathbf{I}_2(\delta_2))$$

or, equivalently, (in case $p_{1,n} > 1$)

$$\tfrac{1}{p_{1,n}-1}(\ln\det(\mathbf{I}^{(n)}(\overline{\delta}_1 \otimes \delta_2)) - p_{2,1}\ln\det(\mathbf{I}_1^{(n)}(\overline{\delta}_1))) = \ln\det(\textstyle\int f_1 f_1'\, d\delta_2) + \tfrac{1}{p_{1,n}-1}\ln\det(\mathbf{I}_2(\delta_2)) .$$

This equality yields

$$\ln\det(\textstyle\int f_1 f_1'\, d\delta_2) \le \ln\det(\int f_1 f_1'\, d\delta^*_{2,n}) + \tfrac{1}{p_{1,n}-1}(\ln\det(\mathbf{I}_2(\delta^*_{2,n})) - \ln\det(\mathbf{I}_2(\delta_2))) .$$

As $\ln\det(\mathbf{I}_2(\cdot))$ is bounded from above by $\gamma = \ln\det(\mathbf{I}_2(\delta_2^*))$, where δ_2^* is D-optimum in the second marginal model, we can conclude that

$$\begin{aligned}\ln\det(\textstyle\int f_1 f_1'\, d\delta_2) &\le \lim_{n\to\infty}\ln\det(\textstyle\int f_1 f_1'\, d\delta^*_{2,n}) + \limsup_{n\to\infty}\tfrac{1}{p_{1,n}-1}(\gamma - \ln\det(\mathbf{I}_2(\delta_2))) \\ &= \ln\det(\textstyle\int f_1 f_1'\, d\widetilde{\delta}_2)\end{aligned}$$

and, hence, δ_2 is dominated by $\widetilde{\delta}_2$, where the last equality follows from $\delta^*_{2,n} \xrightarrow{\mathcal{D}} \widetilde{\delta}_2$ and $p_{1,n} \to \infty$.

Next, we consider the case $\mathbf{1} \notin \mathrm{span}(a_1^{(1,n)}, ..., a_{p_{1,n}}^{(1,n)})$ for every n. From Lemma 6.5 (ii) we obtain

$$\det(\mathbf{I}^{(n)}(\overline{\delta}_1 \otimes \delta_2)) = \det(\mathbf{I}_1^{(n)}(\overline{\delta}_1))^{p_{2,1}} \det(\textstyle\int f_1 f_1'\, d\delta_2)^{p_{1,n}} \det(\int f_2 f_2'\, d\delta_2)$$

or, equivalently,

$$\frac{1}{p_{1,n}}(\ln\det(\mathbf{I}^{(n)}(\overline{\delta}_1\otimes\delta_2))-p_{2,1}\ln\det(\mathbf{I}_1^{(n)}(\overline{\delta}_1)))=\ln\det(\textstyle\int f_1f_1'\,d\delta_2)+\frac{1}{p_{1,n}}\ln\det(\int f_2f_2'\,d\delta_2)\,.$$

By the same arguments as above we get $\ln\det(\int f_1f_1'\,d\delta_2)\le\ln\det(\int f_1f_1'\,d\widetilde{\delta}_2)$, again, for every $\delta_2\in\Delta_2(\beta^{(2)})$.

The general case where $\mathbf{1}\in\text{span}(a_1^{(1,n)},...,a_{p_{1,n}}^{(1,n)})$ for some, but not all n can be treated by investigating subsequences of the pure type considered before.

Now, let $\delta_2\notin\Delta_2(\beta^{(2)})$. Then for every $\varepsilon>0$ there is a design $\delta_2^\varepsilon\in\Delta_2(\beta^{(2)})$ such that

$$\ln\det(\textstyle\int f_1f_1'\,d\delta_2)\le\ln\det(\int f_1f_1'\,d\delta_2^\varepsilon)+\varepsilon\le\ln\det(\int f_1f_1'\,d\widetilde{\delta}_2)+\varepsilon\,.$$

Thus by letting ε tend to zero we obtain $\ln\det(\int f_1f_1'\,d\delta_2)\le\ln\det(\int f_1f_1'\,d\widetilde{\delta}_2)$ and hence, finally, the D-optimality of $\widetilde{\delta}_2$ in the reduced marginal model.

(ii) We proceed as in the proof of part (i). Denote by $\mathbf{C}^{(n)}=\mathbf{I}^{(n)^{-1}}$ and $\mathbf{C}_1^{(n)}=\mathbf{I}_1^{(n)^{-1}}$ the covariance matrices in the nth model and its corresponding marginal model $\mu_1^{(n)}(t_1)=a^{(1,n)}(t_1)'\beta^{(1)}$, $t_1\in T_{1,n}$, respectively. For every $\delta_2\in\Delta_2(\beta^{(2)})$ we obtain

$$\text{tr}(\mathbf{C}^{(n)}(\overline{\delta}_1\otimes\delta_2))=\alpha_n\text{tr}((\textstyle\int f_1f_1'\,d\delta_2)^{-1})+\gamma_n(\overline{\delta}_1)\text{tr}(\mathbf{C}_{2,\beta_1^{(2)}}(\delta_2))+\text{tr}(\mathbf{C}_{2,\beta_2}(\delta_2))$$

by Lemma 6.5, where $\alpha_n=\text{tr}(\mathbf{C}_1^{(n)}(\overline{\delta}_1))-\gamma_n(\overline{\delta}_1)$, or, equivalently, (in case $p_{1,n}>1$)

$$\frac{1}{\alpha_n}\text{tr}(\mathbf{C}^{(n)}(\overline{\delta}_1\otimes\delta_2))=\text{tr}((\textstyle\int f_1f_1'\,d\delta_2)^{-1})+\frac{1}{\alpha_n}(\gamma_n(\overline{\delta}_1)\text{tr}(\mathbf{C}_{2,\beta_1^{(2)}}(\delta_2))+\text{tr}(\mathbf{C}_{2,\beta_2}(\delta_2)))\,.$$

(Note that $\alpha_n>0$ if $p_{1,n}>1$ and that $\gamma_n(\overline{\delta}_1)=0$ for $\mathbf{1}\notin\text{span}(a_1^{(1,n)},...,a_{p_{1,n}}^{(1,n)})$.)

$\text{tr}(\mathbf{C}_{2,\beta_2}(\delta_2))$ and $\text{tr}(\mathbf{C}_{2,\beta_1^{(2)}}(\delta_2))$ are both positive for $\delta_2\in\Delta_2(\beta^{(2)})$ and, thus,

$$\begin{aligned}&\text{tr}((\textstyle\int f_1f_1'\,d\delta_2)^{-1})\\ &\ge\liminf_{n\to\infty}\frac{1}{\alpha_n}\text{tr}(\mathbf{C}^{(n)}(\overline{\delta}_1\otimes\delta_2))-\limsup_{n\to\infty}\frac{1}{\alpha_n}(\gamma_n(\overline{\delta}_1)\text{tr}(\mathbf{C}_{2,\beta_1^{(2)}}(\delta_2))+\text{tr}(\mathbf{C}_{2,\beta_2}(\delta_2)))\\ &\ge\lim_{n\to\infty}\text{tr}((\textstyle\int f_1f_1'\,d\delta_{2,n}^*)^{-1})\\ &=\text{tr}((\textstyle\int f_1f_1'\,d\widetilde{\delta}_2)^{-1})\end{aligned}$$

where the second inequality follows from $\frac{1}{\alpha_n}\gamma_n(\overline{\delta}_1)\to0$, $\frac{1}{\alpha_n}\to0$, and the optimality of $\delta_{2,n}^*$, and the last equality holds because of $\delta_{2,n}^*\xrightarrow{\mathcal{D}}\widetilde{\delta}_2$.

That the relation $\text{tr}((\int f_1f_1'\,d\delta_2)^{-1})\ge\text{tr}((\int f_1f_1'\,d\widetilde{\delta}_2)^{-1})$ is also valid for $\delta_2\notin\Delta(\beta^{(2)})$ can be obtained in the same way as for the D-criterion in the proof of (i). Hence, $\widetilde{\delta}_2$ is A-optimum in the reduced marginal model.

(iii) The sequence $(\delta_{2,n}^*)_{n\in\mathbb{N}}$ of designs regarded as a sequence of probability measures is obviously tight, because T_2 is compact. According to PROHOROV's Theorem (see BILLINGSLEY (1968), p 37) for every subsequence of $(\delta_{2,n}^*)_{n\in\mathbb{N}}$ there is a sub-subsequence which is weakly convergent to some design $\widetilde{\delta}_2$. By (i) $\widetilde{\delta}_2$ is D-optimum (resp. A-optimum by (ii)) in the reduced marginal model and, furthermore, $\widetilde{\delta}_2=\delta_2^*$ by the uniqueness of δ_2^*. Hence for every subsequence of $(\delta_{2,n}^*)_{n\in\mathbb{N}}$ there is a sub-subsequence which converges weakly to δ_2^* which proves the assertion (see BILLINGSLEY (1968), p 16). □

For illustrative purposes we apply Theorem 6.16 to models of type (6.7) in which the first factor is specified by a one-way layout and for which the statements of Theorem 6.16 simplify. A further application is indicated in Example 6.12.

Corollary 6.17

For every $I \geq 1$ consider the two-factor model $\mu^{(I)}(i,u) = f_1(u)'\beta_{1,i} + f_2(u)'\beta_2$, $i = 1,...,I$, $u \in U$. Let $\overline{\delta}_{1,I}$ be the uniform design on $\{1,...,I\}$ and let $\delta_{2,I}^* \in \Delta_2$ such that $\overline{\delta}_{1,I} \otimes \delta_{2,I}^*$ is D- (resp. A-) optimum.

(i) *If $\delta_{2,I}^* \xrightarrow{\mathcal{D}} \widetilde{\delta}_2$, then $\widetilde{\delta}_2$ is D- (resp. A-) optimum in the reduced marginal model $\mu_2^{(1)}(u) = f_1(u)'\beta_1^{(2)}$, $u \in U$.*

(ii) *Conversely, if there is a unique D- (resp. A-)optimum design δ_2^* in the reduced marginal model, then $\delta_{2,I}^* \xrightarrow{\mathcal{D}} \delta_2^*$.*

Proof

Because of $p_{1,I} = I \to \infty$, $\mathrm{tr}(\mathbf{C}_1^I(\overline{\delta}_{1,I})) = I^2 \to \infty$, and $\gamma(\overline{\delta}_{1,I}) = I$ the assumptions of Theorem 6.16 are fulfilled. □

If there is a finite subset $T_2^{(0)}$ of T_2 such that the support of $\delta_{2,n}^*$ is contained in $T_2^{(0)}$ for every $n \in \mathbb{N}$, then the weak convergence of $\delta_{2,n}^*$ to δ_2^* is equivalent to the condition that for every $t_2 \in T_2^{(0)}$ the weights $\delta_{2,n}^*(\{t_2\})$ tend to $\delta_2^*(\{t_2\})$. Note that such a behaviour is exhibited in Table 6.1 of Example 6.8. In practical applications, however, these asymptotic results may be useless because in some situations the limiting design δ_2^* has a singular information matrix in the whole marginal model (see e. g. the entries 4 to 7 for D-optimality resp. 3 to 7 for A-optimality in Table 6.1). Therefore, the product design $\overline{\delta}_{1,n} \otimes \delta_2^*$ in which the optimum marginal design $\delta_{2,n}^*$ is substituted by the limiting design δ_2^* cannot be used and its efficency has to be checked for every particular case.

Example 6.9 *One-way layout with additional regression (cf Example 5.3):*

We are interested in a standard parametrization $\mu(i,u) = \mu_0 + \alpha_i + f(u)'\beta_2$ with a general mean μ_0 for a two-factor model in which the influence of the first factor is described by a one-way layout (with general mean). Thus, we have to impose the identifiability condition $\sum_{i=1}^I \alpha_i = \mathbf{1}_I'\alpha = 0$ which is invariant with respect to permutations of the levels (cf Example 3.8). An appropriate set $\bar{a}^{(1)}$ of regression functions which reflect the identifiability condition is given by $\bar{a}^{(1)}(i) = \mathbf{e}_{I,i} - \frac{1}{I}\mathbf{1}_I$ for the qualitative factor. (Note that the resulting first marginal model $\bar{\mu}_1(i) = \alpha_i$ differs from the common one-way layout because of the additional linear restriction $\mathbf{1}_I'\alpha = 0$.) The present model can be rewritten as $\mu(i,u) = \bar{a}^{(1)}(i)'\alpha + f_2(u)'\overline{\beta}_2$ with $f = (1, f')'$ and identifiability condition $\sum_{i=1}^I \alpha_i = 0$. Hence, with $f_1 = \mathbf{1}$ the model is of the form (6.3) with $\mathbf{1} \notin \mathrm{span}(\bar{a}_1^{(1)}, ..., \bar{a}_I^{(1)})$. (Note that in this case the components of f_1 and f_2 are not linearly independent.) Permutations of the levels induce linear transformations of the regression function $\bar{a}^{(1)}$ by means of the associated permutation matrices and the linear identifiability condition is invariant. Hence, we can find an symmetrized A-optimum design $\overline{\delta}_1 \otimes \delta_2^*$ and the second marginal δ_2^* is A-optimum in the second marginal model $\mu_2(u) = \beta_0 + f(u)'\beta_2$ due to the block diagonal structure of the information matrix $\mathbf{I}(\overline{\delta}_1 \otimes \delta_2^*)$ (see Lemma 6.5 (ii)) in accordance with the result of Theorem 6.13. Note, also, that the regression function $\bar{a}^{(1)}$ is centered with respect to the uniform design $\overline{\delta}_1$ and the present problem is essentially covered by the concept of orthogonal designs (Subsection 5.2) up to identifiability conditions. □

In order to show that the methodology of symmetrization is very helpful also in seemingly more complicated situations we will answer the question posed by KUROTSCHKA

(1988) for an optimum design in the following large class of examples,

Example 6.10 *K-way layout without interactions combined with quadratic regression:*

$$\mu(i_1, ..., i_K, u) = \sum\nolimits_{k=1}^{K} \alpha_{i_k}^{(k)} + \beta_{(i_1,i_2,...,i_M),1} u + \beta_{(i_1,i_2,...,i_M),2} u^2 \ ,$$

$i_k = 1, ..., I_k$, $k = 1, ..., K$, $u \in U$. We assume that the K-way layout part of the model is given in standard parametrization (without general mean), i. e. $\sum_{i_k=1}^{I_k} \alpha_{i_k}^{(k)} = 0$ for $k = 2, ..., K$, but note that the choice of the parametrization is immaterial if we consider the D-criterion. We assume that the second marginal design region U for the quantitative factor is standardized, $U = [-1, 1]$.

A closer look at the model shows that the parameters associated with the quantitative factor specified by u may vary with the level combinations $(i_1, ..., i_M)$ of the first M factors in the K-way layout, $0 \leq M \leq K$, and that there is no interaction with the remaining last $K - M$ factors. In particular, the case $M = 0$ corresponds to the situation of no interaction between the factors of the K-way layout and the quantitative factor and can be treated by the methods of Section 5.

In the present setting we can use an extended regression function $a : \times_{k=1}^{K}\{1, ..., I_k\} \times U \to I\!R^p$, $p = \sum_{k=1}^{K} I_k + 2\prod_{k=1}^{M} I_k$, with linearly dependent components $a(i_1, ..., i_K, u) = (\mathbf{e}'_{I_1,i_1}, ..., \mathbf{e}'_{I_K,i_K}, ((\bigotimes_{k=1}^{M} \mathbf{e}_{I_k,i_k}) \otimes f(u))')'$, where $f(u) = (u, u^2)'$, and linear identifiability conditions $L_0\beta = \mathbf{0}$ given by $L_0 = (\mathbf{0}|\mathrm{diag}((\mathbf{1}'_{I_k})_{k=2,...,K})|\mathbf{0}) \in I\!R^{(K-1)\times p}$. As for any of the K factors of the K-way layout the permutations of the corresponding levels induce linear transformations we may choose the combination of all these permutations $G = \{g = (g_1, ..., g_K, \mathrm{id}|_U);\ g_k \in G_k, k = 1, ..., K\}$ as the transformation group G acting on T, where each G_k is the permutation group on $\{1, ..., I_k\}$ (cf Example 3.3). The associated linear transformation matrices

$$Q_g = \begin{pmatrix} \mathrm{diag}((Q_{k,g_k})_{k=1,...,K}) & \mathbf{0} \\ \mathbf{0} & (\bigotimes_{k=1}^{M} Q_{k,g_k}) \otimes \mathbf{E}_2 \end{pmatrix}$$

are orthogonal because Q_{k,g_k} is a permutation matrix on $\{1, ..., I_k\}$ for every $g_k \in G_k$. Furthermore, the identifiability condition is invariant, $L_0 Q_g = L_0$. We can confine to the essentially complete class of designs which are invariant with respect to G, i. e. to designs $\overline{\delta} = \bigotimes_{k=1}^{K} \overline{\delta}_{1,k} \otimes \delta_2$ where $\overline{\delta}_{1,k}$ is the uniform design on the design region $\{1, ..., I_k\}$ of the kth factor of the K-way layout (see Theorem 3.13), if we are interested in an invariant criterion like the Φ_q-criteria and, in particular, in the D-criterion which will be considered here. It remains to find an optimum marginal design δ_2^* for the quantitative factor which maximizes $\det(\mathbf{I}(\bigotimes_{k=1}^{K} \overline{\delta}_{1,k} \otimes \delta_2))$.

In the case of no interactions ($M = 0$) Corollary 5.18 establishes directly the D-optimality of the product design $\delta^* = \bigotimes_{k=1}^{K} \overline{\delta}_{1,k} \otimes \delta_2^*$ if δ_2^* is D-optimum in the pure quadratic regression model $\mu_2(u) = \beta_0 + \beta_1 u + \beta_2 u^2$, $u \in U$, i. e. if δ_2^* assigns equal weight $\frac{1}{3}$ to -1, 0, and 1 (cf Example 3.9).

In case $M > 0$ we notice first that there is no interaction between the last $K - M$ factors of the K-way layout and the remaining factors and we obtain from Corollary 5.18 that the product design $\delta^* = \bigotimes_{k=1}^{K} \overline{\delta}_{1,k} \otimes \delta_2^*$ is D-optimum if its $(M + 1)$-dimensional marginal $\delta_{(M)}^* = \bigotimes_{k=1}^{M} \overline{\delta}_{1,k} \otimes \delta_2^*$ is D-optimum in the submodel

$$\mu^{((1,...,M),2)}(i_1, ..., i_M, u) = \sum\nolimits_{k=1}^{M} \alpha_{i_k}^{(k)} + \beta_{(i_1,i_2,...,i_M),1} u + \beta_{(i_1,i_2,...,i_M),2} u^2$$

of a M-way layout combined with quadratic regression with identifiability conditions $\sum_{i_k=1}^{I_k} \alpha_{i_k}^{(k)} = 0$ for $k = 2, ..., M$. Thus, the last $K - M$ factors have no impact on δ_2^* and we can confine to the case $K = M$ of the quantitatvie factor interacting with all qualitiative ones in further calculations.

For $M = 1$ the M-way layout combined with quadratic regression reduces to a model $\mu^{((1),2)}(i_1, u) = \alpha_{i_1}^{(1)} + \beta_{(i_1),1}u + \beta_{(i_1),2}u^2$ with complete product-type interactions of the two factors. Hence by Theorem 4.2 the product design $\overline{\delta}_1 \otimes \delta_2^*$ is D-optimum in this model if δ_2^* is D-optimum itself in the pure quadratic regression model, i. e. $\delta_2^* = \frac{1}{3}(\epsilon_1 + \epsilon_{-1} + \epsilon_0)$.

For $M > 1$ the situation is more complicated because of the hidden influence of the presence of identifiability conditions. Denote by $\mathbf{I}^{(m)}$ the information matrix in the m-way layout $\mu^{((1,...,m),2)}(i_1, ..., i_m, u) = \sum_{k=1}^m \alpha_{i_k}^{(k)} + \beta_{(i_1,i_2,...,i_m),1}u + \beta_{(i_1,i_2,...,i_m),2}u^2$ without interactions combined with quadratic regression which is the submodel associated with the first m qualitative factors and the quantitative factor. Furthermore, let $f^{(m)}$ be defined by $f^{(m)}(i_1, ..., i_m, u) = (1, (\bigotimes_{k=1}^m \mathbf{e}'_{I_k,i_k}) \otimes (u, u^2))'$, then $\mu^{((1,...,m),2)}(i_1, ..., i_m, u) = \sum_{k=1}^m \widetilde{\alpha}_{i_k}^{(k)} + f^{(m)}(i_1, i_2, ..., i_m, u)'\beta^{(m)}$ is a reparametrization of that model. As in the proof of Theorem 6.12 we obtain

$$\begin{aligned}
&\det\left(\mathbf{I}^{(m)}(\textstyle\bigotimes_{k=1}^m \overline{\delta}_{1,k} \otimes \delta_2)\right) \\
&= \gamma_m \det\left(\mathbf{I}^{(m-1)}(\textstyle\bigotimes_{k=1}^{m-1} \overline{\delta}_{1,k} \otimes \delta_2)\right) \det\left(\int f^{(m-1)} f^{(m-1)\prime}\, d(\textstyle\bigotimes_{k=1}^{m-1} \overline{\delta}_{1,k} \otimes \delta_2)\right)^{I_m - 1}
\end{aligned}$$

for every $\delta_2 \in \Delta(\beta^{(2)})$, where $\gamma_m > 0$ is independent of δ_2. By repeating this reduction to an $(m-1)$-dimensional submodel we, finally, get

$$\begin{aligned}
&\det\left(\mathbf{I}^{(M)}(\textstyle\bigotimes_{k=1}^M \overline{\delta}_{1,k} \otimes \delta_2)\right) \\
&= \gamma \det(\mathbf{I}_2(\delta_2))^{I_1} \textstyle\prod_{m=1}^{M-1} \det\left(\int f_1^{(m)} f_1^{(m)\prime}\, d(\textstyle\bigotimes_{k=1}^m \overline{\delta}_{1,k} \otimes \delta_2)\right)^{I_{m+1}-1}
\end{aligned}$$

for some $\gamma > 0$. Next, we obtain for the reduced information matrices

$$\begin{aligned}
&\int f_1^{(m)} f_1^{(m)\prime}\, d(\textstyle\bigotimes_{k=1}^m \overline{\delta}_{1,k} \otimes \delta_2) \\
&= \begin{pmatrix} 1 & \frac{1}{I_1 \cdot \ldots \cdot I_m} \mathbf{1}'_{I_1\cdot\ldots\cdot I_m} \otimes \begin{pmatrix} m_1(\delta_2) & m_2(\delta_2) \end{pmatrix} \\ \frac{1}{I_1 \cdot \ldots \cdot I_m} \mathbf{1}_{I_1\cdot\ldots\cdot I_m} \otimes \begin{pmatrix} m_1(\delta_1) \\ m_2(\delta_2) \end{pmatrix} & \frac{1}{I_1 \cdot \ldots \cdot I_m} \mathbf{E}_{I_1\cdot\ldots\cdot I_m} \otimes \begin{pmatrix} m_2(\delta_2) & m_3(\delta_2) \\ m_3(\delta_2) & m_4(\delta_2) \end{pmatrix} \end{pmatrix},
\end{aligned}$$

where $m_j(\delta_2) = \int u^j\, \delta_2(du)$, $j = 1, 2, ...$ With the same arguments as used in Example 6.8 we can confine to those designs $\delta_2 = \delta_2^{(w_2)}$ which assign equal weights w_2 to each of the endpoints -1 and 1 of the interval U and weight $1 - 2w_2$ to the midpoint 0. Hence, $m_j(\delta_2^{(w_2)}) = 0$ for j odd and $m_4(\delta_2^{(w_2)}) = m_2(\delta_2^{(w_2)}) = 2w_2$. By the formula for the determinant of partitioned matrices (Lemma A.2) we get

$$\det\left(\int f_1^{(m)} f_1^{(m)\prime}\, d(\textstyle\bigotimes_{k=1}^m \overline{\delta}_{1,k} \otimes \delta_2)\right) = \bar{\gamma}_m (1 - 2w_2)(2w_2)^{2\prod_{k=1}^m I_k}$$

for some $\bar{\gamma}_m > 0$ which is independent of w_2. Combining all these results we obtain

$$\det\left(\mathbf{I}^{(M)}(\textstyle\bigotimes_{k=1}^M \overline{\delta}_{1,k} \otimes \delta_2^{(w_2)})\right) = \bar{\gamma}(2w_2)^{2\prod_{k=1}^M I_k}(1 - 2w_2)^{\sum_{k=1}^M I_k - (M-1)}$$

where $\bar{\gamma} > 0$. Then the optimum weight is $w_2^* = p_{(M)}^{-1} \prod_{k=1}^M I_k$, where $p_{(M)} = 2\prod_{k=1}^M I_k + \sum_{k=1}^M I_k - (M-1)$ is the number of parameters in the M-way layout combined with quadratic regression.

Thus, we can conclude that the product design $\delta^* = \bigotimes_{k=1}^K \overline{\delta}_{1,k} \otimes \delta_2^{(w_2^*)}$ is D-optimum in the original K-way layout combined with quadratic regression. (Note that $w_2^* \to \frac{1}{2}$ as M increases.) □

By the ideas pointed out in Example 6.10 even more complicated models can be considered, e. g. the situation that the regression part is a polynomial of a higher degree than. Howevr, as in the pure regression situation (see KARLIN and STUDDEN (1966)) the design points are roots of higher order polynomials which cannot be explicitly solved, in general. For another large class of examples we look at the effect of different interaction structures between two qualitative factors and one quantitative factor,

Example 6.11 *Two-way layout combined with quadratic regression:*
In Table 6.2 we have collected a variety of three-factor models $\mu(i,j,u)$ where the marginal models of the first and second factor are one-way layouts $\mu_1(i) = \alpha_i^{(1)}$, $i = 1..., I$, and $\mu_2(j) = \alpha_j^{(2)}$, $j = 1..., J$, respectively, and the third factor is modeled by a quadratic regression $\mu_3(u) = \beta_0^{(3)} + \beta_1^{(3)}u + \beta_2^{(3)}u^2$, $u \in U = [-1,1]$, with standardized design region U. Together with the models of Table 6.1 in which at most one of the qualitative factors is active a complete list of possible models is presented (up to symmetries).

Because the considerations of Example 6.10 carry over to all models in Table 6.2 there will be always a D-optimum product design $\delta^* = \overline{\delta}_1 \otimes \overline{\delta}_2 \otimes \delta_3^{(w)}$ where $\overline{\delta}_1$ and $\overline{\delta}_2$ are the uniform designs on $\{1,...,I\}$ and $\{1,...,J\}$ respectively, and $\delta_3^{(w)}$ assigns weights w to each of the endpoints -1 and 1 of the interval U and the remaining weight $1-2w$ to the midpoint 0. Hence, only the optimum weights w^* and $1-2w^*$ may vary with the different interaction structures, and we have listed the latter in Table 6.2. A related example has been considered by LIM, STUDDEN, and WYNN (1988). □

	$\mu(i,j,u)$	$1-2w^*$
1	$\alpha_i^{(1)} + \alpha_j^{(2)} + \beta_1 u + \beta_2 u^2$	$\frac{1}{3}$
2	$\alpha_i^{(1)} + \alpha_j^{(2)} + \beta_{1i} u + \beta_{2i} u^2$	$\frac{1}{3}$
3	$\alpha_i^{(1)} + \alpha_j^{(2)} + \beta_{1i}^{(1)} u + \beta_{1j}^{(2)} u + \beta_{2i}^{(1)} u^2 + \beta_{2j}^{(2)} u^2$	$\frac{1}{3}$
4	$\alpha_i^{(1)} + \alpha_j^{(2)} + \beta_{1ij} u + \beta_{2ij} u^2$	$\frac{I+J-1}{2IJ+I+J-1}$
5	$\alpha_i^{(1)} + \alpha_j^{(2)} + \beta_{1i}^{(1)} u + \beta_{1j}^{(2)} u + \beta_{2ij} u^2$	$\frac{I+J-1}{IJ+2(I+J-1)}$

Table 6.2: Two-way layout combined with quadratic regression, optimal weights at 0

$\mu(i,j,u)$	$1-2w^*$
6 $\alpha_i^{(1)}+\alpha_j^{(2)}+\beta_{1ij}u+\beta_{2i}^{(1)}u^2+\beta_{2j}^{(2)}u^2$	$\frac{I+J-1}{IJ+2(I+J-1)}$
7 $\alpha_i^{(1)}+\alpha_j^{(2)}+\beta_{1i}^{(1)}u+\beta_{1j}^{(2)}u+\beta_2u^2$	$\frac{1}{I+J+1}$
8 $\alpha_i^{(1)}+\alpha_j^{(2)}+\beta_1u+\beta_{2i}^{(1)}u^2+\beta_{2j}^{(2)}u^2$	$\frac{I+J-1}{2I+2J-1}$
9 $\alpha_i^{(1)}+\alpha_j^{(2)}+\beta_{1ij}u+\beta_2u^2$	$\frac{1}{IJ+2}$
10 $\alpha_i^{(1)}+\alpha_j^{(2)}+\beta_1u+\beta_{2ij}u^2$	$\frac{I+J-1}{IJ+I+J}$
11 $\alpha_i^{(1)}+\alpha_j^{(2)}+\beta_{1i}^{(1)}u+\beta_{1j}^{(2)}u+\beta_{2i}u^2$	$\frac{I}{3I+J-1}$
12 $\alpha_i^{(1)}+\alpha_j^{(2)}+\beta_{1i}u+\beta_{2i}^{(1)}u^2+\beta_{2j}^{(2)}u^2$	$\frac{I+J-1}{3I+2J-2}$
13 $\alpha_i^{(1)}+\alpha_j^{(2)}+\beta_{1i}u+\beta_{2ij}u^2$	$\frac{I+J-1}{IJ+2I+J-1}$
14 $\alpha_i^{(1)}+\alpha_j^{(2)}+\beta_{1ij}u+\beta_{2i}u^2$	$\frac{1}{J+2}$
15 $\alpha_i^{(1)}+\alpha_j^{(2)}+\beta_{1i}u+\beta_2u^2$	$\frac{1}{I+2}$
16 $\alpha_i^{(1)}+\alpha_j^{(2)}+\beta_1u+\beta_{2i}u^2$	$\frac{I}{2I+1}$
17 $\alpha_i^{(1)}+\alpha_j^{(2)}+\beta_{1i}u+\beta_{2j}u^2$	$\frac{J}{I+2J}$
18 $\alpha_{ij}+\beta_1u+\beta_2u^2$	$\frac{1}{3}$
19 $\alpha_{ij}+\beta_{1i}u+\beta_{2i}u^2$	$\frac{1}{3}$
20 $\alpha_{ij}+\beta_{1ij}u+\beta_{2ij}u^2$	$\frac{1}{3}$
21 $\alpha_{ij}+\beta_{1i}^{(1)}u+\beta_{1j}^{(2)}u+\beta_{2i}^{(1)}u^2+\beta_{2j}^{(2)}u^2$	$\frac{1}{3}$

Table 6.2: *continued*

	$\mu(i,j,u)$	$1-2w^*$
22	$\alpha_{ij}+\beta_{1ij}u+\beta_{2i}^{(1)}u^2+\beta_{2j}^{(2)}u^2$	$\dfrac{I+J-1}{IJ+2(I+J-1)}$
23	$\alpha_{ij}+\beta_{1i}^{(1)}u+\beta_{1j}^{(2)}u+\beta_{2ij}u^2$	$\dfrac{IJ}{2IJ+I+J-1}$
24	$\alpha_{ij}+\beta_{1ij}u+\beta_2u^2$	$\dfrac{1}{IJ+2}$
25	$\alpha_{ij}+\beta_1u+\beta_{2ij}u^2$	$\dfrac{IJ}{2IJ+1}$
26	$\alpha_{ij}+\beta_{1i}^{(1)}u+\beta_{1j}^{(2)}u+\beta_2u^2$	$\dfrac{1}{I+J+1}$
27	$\alpha_{ij}+\beta_1u+\beta_{2i}^{(1)}u^2+\beta_{2j}^{(2)}u^2$	$\dfrac{I+J-1}{2I+2J-1}$
28	$\alpha_{ij}+\beta_{1ij}u+\beta_{2i}u^2$	$\dfrac{1}{J+2}$
29	$\alpha_{ij}+\beta_{1i}u+\beta_{2ij}u^2$	$\dfrac{J}{2J+1}$
30	$\alpha_{ij}+\beta_{1i}u+\beta_{2i}^{(1)}u^2+\beta_{2j}^{(2)}u^2$	$\dfrac{I+J-1}{3I+2J-2}$
31	$\alpha_{ij}+\beta_{1i}^{(1)}u+\beta_{1j}^{(2)}u+\beta_{2i}u^2$	$\dfrac{I}{3I+J-1}$
32	$\alpha_{ij}+\beta_{1i}u+\beta_2u^2$	$\dfrac{1}{I+2}$
33	$\alpha_{ij}+\beta_1u+\beta_{2i}u^2$	$\dfrac{I}{2I+1}$
34	$\alpha_{ij}+\beta_{1i}u+\beta_{2j}u^2$	$\dfrac{J}{I+2J}$
35	$\alpha_i+\beta_{1i}^{(1)}u+\beta_{1j}^{(2)}u+\beta_{2i}^{(1)}u^2+\beta_{2j}^{(2)}u^2$	$\dfrac{I}{3I+2J-2}$
36	$\alpha_i+\beta_{1ij}u+\beta_{2ij}u^2$	$\dfrac{1}{2J+1}$
37	$\alpha_i+\beta_{1j}u+\beta_{2j}u^2$	$\dfrac{1}{2J+1}$

Table 6.2: *continued*

$\mu(i,j,u)$	$1-2w^*$
38 $\alpha_i + \beta_{1i}u + \beta_{2i}^{(1)}u^2 + \beta_{2j}^{(2)}u^2$	$\frac{I}{3I+J-1}$
39 $\alpha_i + \beta_{1i}^{(1)}u + \beta_{1j}^{(2)}u + \beta_{2i}u^2$	$\frac{I}{3I+J-1}$
40 $\alpha_i + \beta_{1i}u + \beta_{2ij}u^2$	$\frac{1}{J+2}$
41 $\alpha_i + \beta_{1ij}u + \beta_{2i}u^2$	$\frac{1}{J+2}$
42 $\alpha_i + \beta_{1i}u + \beta_{2j}u^2$	$\frac{1}{I+J+1}$
43 $\alpha_i + \beta_{1j}u + \beta_{2i}u^2$	$\frac{I}{2I+J}$
44 $\alpha_i + \beta_{1i}^{(1)}u + \beta_{1j}^{(2)}u + \beta_2u^2$	$\frac{1}{I+J+1}$
45 $\alpha_i + \beta_1u + \beta_{2i}^{(1)}u^2 + \beta_{2j}^{(2)}u^2$	$\frac{I}{2I+J}$
46 $\alpha_i + \beta_{1i}^{(1)}u + \beta_{1j}^{(2)}u + \beta_{2ij}u^2$	$\frac{I}{IJ+2I+J-1}$
47 $\alpha_i + \beta_{1ij}u + \beta_{2i}^{(1)}u^2 + \beta_{2j}^{(2)}u^2$	$\frac{I}{IJ+2I+J-1}$
48 $\alpha_i + \beta_{1ij}u + \beta_2u^2$	$\frac{1}{IJ+2}$
49 $\alpha_i + \beta_1u + \beta_{2ij}u^2$	$\frac{I}{IJ+I+1}$
50 $\alpha_i + \beta_{1i}^{(1)}u + \beta_{1j}^{(2)}u + \beta_{2j}u^2$	$\frac{1}{I+2J}$
51 $\alpha_i + \beta_{1j}u + \beta_{2i}^{(1)}u^2 + \beta_{2j}^{(2)}u^2$	$\frac{I}{2I+2J-1}$
52 $\alpha_i + \beta_{1ij}u + \beta_{2j}u^2$	$\frac{1}{IJ+J+1}$
53 $\alpha_i + \beta_{1j}u + \beta_{2ij}u^2$	$\frac{I}{IJ+I+J}$

Table 6.2: *continued*

$\mu(i,j,u)$	$1-2w^*$
54 $\alpha_i+\beta_1 u+\beta_{2j}u^2$	$\frac{1}{J+2}$
55 $\alpha_i+\beta_{1j}u+\beta_2 u^2$	$\frac{1}{J+2}$
56 $\alpha_0+\beta_{1i}^{(1)}u+\beta_{1j}^{(2)}u+\beta_{2i}^{(1)}u^2+\beta_{2j}^{(2)}u^2$	$\frac{1}{2I+2J-1}$
57 $\alpha_0+\beta_{1ij}u+\beta_{2ij}u^2$	$\frac{1}{2IJ+1}$
58 $\alpha_0+\beta_{1i}^{(1)}u+\beta_{1j}^{(2)}u+\beta_{2ij}u^2$	$\frac{1}{IJ+I+J}$
59 $\alpha_0+\beta_{1ij}u+\beta_{2i}^{(1)}u^2+\beta_{2j}^{(2)}u^2$	$\frac{1}{IJ+I+J}$
60 $\alpha_0+\beta_{1i}^{(1)}u+\beta_{1j}^{(2)}u+\beta_2 u^2$	$\frac{1}{I+J+1}$
61 $\alpha_0+\beta_1 u+\beta_{2i}^{(1)}u^2+\beta_{2j}^{(2)}u^2$	$\frac{1}{I+J+1}$
62 $\alpha_0+\beta_{1ij}u+\beta_2 u^2$	$\frac{1}{IJ+2}$
63 $\alpha_0+\beta_1 u+\beta_{2ij}u^2$	$\frac{1}{IJ+2}$
64 $\alpha_0+\beta_{1i}^{(1)}u+\beta_{1j}^{(2)}u+\beta_{2i}u^2$	$\frac{1}{2I+J}$
65 $\alpha_0+\beta_{1i}u+\beta_{2i}^{(1)}u^2+\beta_{2j}^{(2)}u^2$	$\frac{1}{2I+J}$
66 $\alpha_0+\beta_{1ij}u+\beta_{2i}u^2$	$\frac{1}{IJ+I+1}$
67 $\alpha_0+\beta_{1i}u+\beta_{2ij}u^2$	$\frac{1}{IJ+I+1}$
68 $\alpha_0+\beta_{1i}u+\beta_{2j}u^2$	$\frac{1}{I+J+1}$

Table 6.2: *continued*

As mentioned earlier the finiteness of the group G_1 of transformations acting on T_1 can be replaced by the requirement that G_1 is a compact group (cf Example 3.10). Then the natural extension $G=\{(g_1,\mathrm{id}|_{T_2});\ g_1\in G_1\}$ is again a compact group. If additionally G_1 acts transitively on T_1, then the uniform distribution $\overline{\delta}$ on T_1 is the unique invariant design. In analogy to Lemma 6.4 we can obtain a characterization of the invariant designs δ on $T = T_1 \times T_2$. A design $\delta \in \Delta$ is invariant with respect to G if and only if $\delta = \overline{\delta}_1 \otimes \delta_2$ for some $\delta_2 \in \Delta_2$. All results of this section remain true if the finiteness of G_1 is replaced by the requirement of G_1 being a compact group and some additional regularity conditions are imposed. We will illustrate the implications of these considerations by the following example,

Example 6.12 *Trigonometric regression combined with a second factor:*

In the general two-factor model $\mu(t_1, t_2) = (a^{(1)}(t_1) \otimes f_1(t_2))'\beta_1 + f_2(t_2)'\beta_2$, $(t_1, t_2) \in T_1 \times T_2$, we specify the first marginal $\mu_1(t_1) = a^{(1)}(t_1)'\beta^{(1)}$ by a trigonometric regression of degree M: $a^{(1)}(t_1) = (1, \sin(t_1), \cos(t_1), ..., \sin(Mt_1), \cos(Mt_1))'$, $t_1 \in T_1 = [0, 2\pi)$, $\beta^{(1)} \in \mathbb{R}^{p_1}$ with $p_1 = 2M + 1$ (see Example 3.10). The group $G_1 = \{g_{1,y};\, g_{1,y}(t_1) = t_1 - y \bmod 2\pi, y \in [0, 2\pi)\}$, acting on T_1 is the group of translations modulo 2π. The group G_1 acts transitively on T_1 and is orthogonal for $a^{(1)}$. The uniform distribution $\overline{\delta}_1 = \frac{1}{2\pi}\lambda|_{T_1}$ on T_1 may be replaced by a discrete uniform design $\delta_1^{(N,\tau)} = \frac{1}{N}\sum_{n=1}^{N} \epsilon_{\tau + \frac{n-1}{N}2\pi}$ on an equidistant grid with at least $p_1 = 2M + 1$ supporting points (see Example 3.10).

Now, let $G = \{g = (g_{1,y}, \mathrm{id}|_{T_2});\, g_{1,y} \in G_1\}$ be the corresponding group of transformations acting on $T = T_1 \times T_2$. Analogously to Lemma 6.4 the invariant designs $\overline{\delta}$ with respect to G are characterized as $\overline{\delta} = \overline{\delta}_1 \otimes \delta_2$. In view of Lemma 6.5 and because of $\int a^{(1)}\, d\overline{\delta}_1 = \int a^{(1)}\, d\delta_1^{(N,\tau)} = \mathbf{e}_{p_1,1}, N \geq p_1$, we notice that $\mathbf{I}(\overline{\delta}_1 \otimes \delta_2) = \mathbf{I}(\delta_1^{(N,\tau)} \otimes \delta_2)$ for every $0 \leq \tau < \frac{1}{N}2\pi, N \in \mathbb{N}$, and every $\delta_2 \in \Delta_2$. Hence, we can replace the uniform design $\overline{\delta}_1$ by a design $\delta_1^{(N,\tau)}$ supported on a fintite grid also in the present situation of a two-factor model. According to an analogue of Theorem 6.12 we obtain that the product design $\delta^* = \overline{\delta}_1 \otimes \delta_2^*$ (and, hence, also $\delta^* = \delta_1^{(N,\tau)} \otimes \delta_2^*$, $N \geq 2M + 1$) is D-optimum if δ_2^* maximizes $\det(\mathbf{I}_2(\delta_2)) \det(\int f_1 f_1'\, d\delta_2)^{2M}$ in $\delta_2 \in \Delta_2(\beta^{(2)})$ resp. A-optimum if δ_2^* minimizes $4M\mathrm{tr}((\int f_1 f_1'\, d\delta_2)^{-1}) + \mathrm{tr}(\mathbf{C}_2(\delta_2))$ in $\delta_2 \in \Delta_2(\beta^{(2)})$.

For the limiting case $M \to \infty$ we observe $p_{1,M} = 2M + 1 \to \infty$, $\mathrm{tr}(\mathbf{C}_1^{(M)}(\overline{\delta}_1)) = 4M + 1 \to \infty$, and $\gamma_M(\overline{\delta}_1)/\mathrm{tr}(\mathbf{C}_1^{(M)}(\overline{\delta}_1)) = \frac{1}{4M+1} \to 0$. Hence any limiting design $\widetilde{\delta}_2$ of the designs $\delta_{2,M}^*$ constituting a sequence of D- (resp. A-) optimum designs $\overline{\delta}_1 \otimes \delta_{2,M}^*$ will be D- (resp. A-) optimum in the reduced marginal model $\mu_2^{(1)}(t_2) = f_1(t_2)'\beta_1^{(2)}$, $t_2 \in T_2$, by an analogue to Theorem 6.16.

For the present model it is possible to determine the eigenvalues of the information matrix and, hence, of the covariance matrix explicitly. Let $\lambda_{2,1}(\delta_2) \leq ... \leq \lambda_{2,p_2}(\delta_2)$ be the eigenvalues of $\mathbf{I}_2(\delta_2)$ and let $\lambda_{2,1}^{(1)}(\delta_2) \leq \cdots \leq \lambda_{2,p_{2,1}}^{(1)}(\delta_2)$ be the eigenvalues of $\int f_1 f_1'\, d\delta_2$. Then the resulting eigenvalues $\frac{1}{2}\lambda_{2,1}^{(1)}(\delta_2), ..., \frac{1}{2}\lambda_{2,p_{2,1}}^{(1)}(\delta_2)$ occur $2M$ times and the eigenvalues $\lambda_{2,1}(\delta_2), ..., \lambda_{2,p_2}(\delta_2)$ occur once in the information matrix

$$\mathbf{I}(\overline{\delta}_1 \otimes \delta_2) = \begin{pmatrix} \int f_1 f_1'\, d\delta_2 & \mathbf{0} & \int f_1 f_2'\, d\delta_2 \\ \mathbf{0} & \frac{1}{2}\mathbf{E}_{2M} \otimes \int f_1 f_1' d\delta_2 & \mathbf{0} \\ \int f_2 f_1'\, d\delta_2 & \mathbf{0} & \int f_2 f_2'\, d\delta_2 \end{pmatrix}.$$

Hence, for increasing degree M the limiting result can be extended to every Φ_q-criterion, $0 \leq q < \infty$, i.e. the limiting design $\widetilde{\delta}_2$ (with $\delta_{2,M}^* \xrightarrow{\mathcal{D}} \widetilde{\delta}_2$) is Φ_q-optimum in the reduced marginal model $\mu_2^{(1)}(t_2) = f_1(t_2)'\beta_1^{(2)}, t_2 \in T_2$. In case $\lambda_{2,1}^{(1)}(\widetilde{\delta}_2) \leq 2\lambda_{2,1}(\widetilde{\delta}_2)$ this result is also valid for the E-criterion ($q = \infty$).

We add that we can use an alternative representation $\mu(t_1, t_2) = (\bar{a}^{(1)}(t_1) \otimes f_1(t_2))'\bar{\beta}_1 + f_1(t_2)'\beta_{1,0} + f_2(t_2)'\beta_2$ of the response function μ by a trigonometric regression function $\bar{a}^{(1)}(t_1) = (\sin(t_1), \cos(t_1), ..., \sin(Mt_1), \cos(Mt_1))'$ without a constant term for the first marginal, where $\bar{f}_2 = \binom{f_1}{f_2}$ collects all the regression functions of the second marginal which do not interact with the first factor. Then the above results can also be obtained by the analogue considerations for the case $\mathbf{1} \notin \text{span}(\bar{a}_1^{(1)}, ..., \bar{a}_{2M}^{(1)})$. □

Concluding remark. In models in which one factor is invariant with respect to a group of transformations transitively acting on the associated marginal design region optimum design can be generated as products in which the marginal is uniform for the invariant factor. While this concept has been implicitly be used for long in analysis of variance models with purely qualitative factors the general approach was first mentioned in KUROTSCHKA, SCHWABE and WIERICH (1992) anticipating the present development of the theoretical background and the variety of applications.

7 Some Additional Results

If the structure of the underlying multi-factor model is more complicated, then there is no good reason to expect product designs to be optimum. Here we will sketch the idea of conditional designs in the situation of one qualitative factor completely interacting with the second factor. These complete interaction structures will include the dependence of the conditional design region $T_2(t_1)$ and the regression function $a^{(2|t_1)}$ on t_1. Some hints on partial interaction structures will be given and, finally, some limitations of the methodology will be made evident.

We start with a model in which a qualitative factor is completely interacting with a second factor

$$\mu(i,u) = f_i(u)'\beta_i \ , \tag{7.1}$$

$u \in U_i$, $i = 1, ..., I$, where the regression function f_i and the design region U_i of the second factor may depend on the level of the first factor. These models are known as general *intra-class models* (cf Example 4.3, SEARLE (1971), pp 355, or BUONACCORSI and IYER (1986)) if the second factor is quantitative and as *nested* or *hierarchical models* (cf SEARLE (1971), pp 249) if the second factor is also qualitative.

In case of the design regions $U_i = U$ independent of the level i of the first factor, A- and D-optimum designs have been obtained by KUROTSCHKA (1984) by a mixture of the optimum designs for the conditional models. This result can directly be extended to design regions U_i depending on the level i and to any Φ_q-criterion, $0 \le q \le \infty$, including the E-criterion besides the A- and D-criterion already mentioned.

To present a concise formulation of the result we begin with some remarks and some additional notations. The design regions U_i of the *conditional models* for the second factor

$$\mu_{2|i}(u) = f_i(u)'\beta_i \ , \tag{7.2}$$

$u \in U_i$, are the *i-cuts* $U_i = \{u;\ (i,u) \in T\}$ of the design region T, hence $T = \bigcup_{i=1}^{I} \{i\} \times U_i$. Similarly, every design δ determines a conditional marginal design $\delta_{2|i}$, defined by $\delta_{2|i}(\cdot) = \frac{1}{\delta_1(\{i\})}\delta(\{i\} \times \cdot)$, on the design region U_i of the ith conditional model, if for its first marginal $\delta_1(\{i\}) > 0$. Conversely, for every set of designs $\delta_{2|i}$ on U_i, $i = 1, ..., I$, and every marginal design δ_1 on $\{1, ..., I\}$ we obtain a design δ on T by a δ_1-*mixture* of $(\delta_{2|i})_{i=1,...,I}$, i.e. by $\delta(\{i\} \times \cdot) = \delta_1(\{i\})\delta_{2|i}(\cdot)$. If all the U_i are equal to each other or if they are embedded in a common design region, then the above decomposition property of δ can be recognized as the desintegration of the design δ into its first marginal δ_1 and the MARKOV kernel (or conditional probability measure) δ_2 with the notation $\delta_2(i, \cdot) = \delta_{2|i}(\cdot)$. Recall that $\beta_i \in \mathbb{R}^{p_2(i)}$ and $p = \sum_{i=1}^{I} p_2(i)$. In addition we denote the information and covariance matrices in the ith conditional model by $\mathbf{I}_{2|i}(\delta_{2|i}) = \int a^{(2|i)} a^{(2|i)'}\, d\delta_{2|i}$ resp. $\mathbf{C}_{2|i}(\delta_{2|i})$, where $a^{(2|i)} = f_i$.

Theorem 7.1

δ^* is Φ_q-optimum in the two factor model $\mu(i,u) = f_i(u)'\beta_i$, $u \in U_i$, $i = 1, ..., I$, if and only if

(i) *the conditional design $\delta^*_{2|i}$ of δ^* is Φ_q-optimum in the ith conditional model $\mu_{2|i}(u) = f_i(u)'\beta_i$, $u \in U_i$, for every $i = 1, ..., I$; and*

(ii) *the weights $\delta^*_1(\{i\})$ of the marginal design δ^*_1 for the first factor satisfy*

$$\delta_1^*(\{i\}) = \frac{\left(\sum_{j=1}^{p_2(i)} \lambda_j\left(\mathbf{C}_{2|i}(\delta_{2|i}^*)\right)^q\right)^{1/(q+1)}}{\sum_{\bar{i}=1}^{I}\left(\sum_{j=1}^{p_2(\bar{i})} \lambda_j\left(\mathbf{C}_{2|\bar{i}}(\delta_{2|\bar{i}}^*)\right)^q\right)^{1/(q+1)}},$$

for $0 < q < \infty$, *resp.*

$$\delta_1^*(\{i\}) = \frac{p_2(i)}{\sum_{\bar{i}=1}^{I} p_2(\bar{i})}$$

for the limiting case of D-optimality ($q = 0$) *or*

$$\delta_1^*(\{i\}) = \frac{\lambda_{\max}\left(\mathbf{C}_{2|i}(\delta_{2|i}^*)\right)}{\sum_{\bar{i}=1}^{I} \lambda_{\max}\left(\mathbf{C}_{2|\bar{i}}(\delta_{2|\bar{i}}^*)\right)}$$

for the limiting case of E-optimality ($q = \infty$).

Proof

First we notice that $\mathbf{I}(\delta) = \mathrm{diag}((\delta_1(\{i\})\mathbf{I}_{2|i}(\delta_{2|i}))_{i=1,...,I})$ is block diagonal for every δ and that $\mathbf{I}(\delta)$ is regular if and only if $\delta_1(\{i\}) > 0$, $i = 1, ..., I$, and each of the conditional information matrices $\mathbf{I}_{2|i}(\delta_{2|i})$ is regular.

Consider the class of designs δ which are associated with a fixed marginal $\delta_1 \in \Delta_1(\beta^{(1)})$. Then a design δ is Φ_q-optimum within this class if and only if the conditional designs $\delta_{2|i}$ are Φ_q-optimum in the conditional models by the block diagonal structure of $\mathbf{I}(\delta)$.

Moreover, for $0 < q < \infty$, the quantity $\sum_{j=1}^{p} \lambda_j(\mathbf{C}(\delta))^q$ related to the *q-norm* of the vector of eigenvalues $(\lambda_j(\mathbf{C}(\delta)))_{j=1,...,p}$ can be calculated according to

$$\sum_{j=1}^{p} \lambda_j(\mathbf{C}(\delta))^q = \sum_{i=1}^{I} \delta_1(\{i\})^{-q} \sum_{j=1}^{p_2(i)} \lambda_j(\mathbf{C}_{2|i}(\delta_{2|i}))^q .$$

By inserting the optimum conditional designs $\delta_{2|i}^*$ and optimizing with respect to the marginal weights $(\delta_1(\{i\}))_{i=1,...,I}$ for the first factor we obtain the claimed equivalence for $0 < q < \infty$.

For the limiting cases $q = 0$ and $q = \infty$ we notice that

$$\det(\mathbf{C}(\delta)) = \prod_{i=1}^{I} \delta_1(\{i\})^{-p_2(i)} \det(\mathbf{C}_{2|i}(\delta_{2|i}))$$

and

$$\lambda_{\max}(\mathbf{C}(\delta)) = \max_{i=1,...,I}(\delta_1(\{i\})^{-1}\lambda_{\max}(\mathbf{C}_{2|i}(\delta_{2|i}))) ,$$

respectively. The unique solutions of the corresponding optimization problem establish the representation of δ^*. □

In particular, for the *A*-criterion the optimum marginal design δ_1^* is given by

$$\delta_1^*(\{i\}) = \frac{\sqrt{\mathrm{tr}(\mathbf{C}_{2|i}(\delta_{2|i}^*))}}{\sum_{\bar{i}=1}^{I} \sqrt{\mathrm{tr}(\mathbf{C}_{2|\bar{i}}(\delta_{2|\bar{i}}^*))}}$$

(cf KUROTSCHKA (1984), Theorem 2.2).

Example 7.1 *One-way layout combined with polynomial regression of differing order:* We consider an intra-class model with a qualitative factor which may be adjusted to two levels and with conditional polynomial regression models associated with these levels which are linear resp. quadratic,

$$\begin{aligned} \mu(1,u) &= \beta_{10} + \beta_{11}u \ , \\ \mu(2,u) &= \beta_{20} + \beta_{21}u + \beta_{22}u^2 \ , \end{aligned}$$

$u \in U_i$, $i = 1,2$. Additionally, we assume that the design regions are standardized, $U_1 = U_2 = [-1,1]$.

In Example 3.9 we have obtained the conditionally A-, D- and E- optimum designs for both linear and quadratic regression. We recall that the design $\delta^*_{2|1}$ which assigns equal weights $\frac{1}{2}$ to the endpoints -1 and 1 of the interval is simultaneously Φ_q-optimum in the conditional model of linear regression. In quadratic regression the optimum design $\delta^*_{2|2}$ assigns weights w^* to the endpoints -1 and 1 and the remaining weight $1-2w^*$ to the midpoint 0 of the interval. This optimum weight w^* equals $\frac{1}{3}$ for the D-optimality, $\frac{1}{4}$ for the A-optimality and $\frac{1}{5}$ for the E-optimality, respectively.

By Theorem 7.1 we can determine the optimum designs for the given intra-class model. As $p_2(1) = 2$ and $p_2(2) = 3$ the optimum marginal design is given by $\delta_1^*(\{1\}) = \frac{2}{5}$ and $\delta_1^*(\{2\}) = \frac{3}{5}$ with respect to the D-criterion. Hence, the D-optimum design assigns equal weights $\frac{1}{5}$ to each of the supporting points $(1,-1)$, $(1,1)$, $(2,-1)$, $(2,0)$ and $(2,1)$ generated from the conditionally optimum designs. (Note that the resulting design has minimum support and the equal weights follow also from Lemma 2.11.) For the A-optimality we calculate $\mathrm{tr}(\mathbf{C}_{2|1}(\delta^*_{2|1})) = 2$ and $\mathrm{tr}(\mathbf{C}_{2|2}(\delta^*_{2|2})) = 8$ and obtain the optimum marginal design as $\delta_1^*(\{1\}) = \frac{1}{3}$ and $\delta_1^*(\{2\}) = \frac{2}{3}$. Hence, the A-optimum design assigns equal weights $\frac{1}{6}$ to each of the endpoints $(1,-1)$, $(1,1)$, $(2,-1)$ and $(2,1)$ of the conditional intervals and the remaining weight $\frac{1}{3}$ to the midpoint $(2,0)$ in the quadratic conditional model. Finally, for the E-optimality we have $\lambda_{\max}(\mathbf{C}_{2|1}(\delta^*_{2|1})) = 1$ and $\lambda_{\max}(\mathbf{C}_{2|2}(\delta^*_{2|2})) = 5$ and the optimum marginal design is given by $\delta_1^*(\{1\}) = \frac{1}{6}$ and $\delta_1^*(\{2\}) = \frac{5}{6}$. Hence, the E-optimum design assigns weights $\frac{1}{12}$ to each of the endpoints $(1,-1)$ and $(1,1)$ in the conditional linear model, weights $\frac{1}{6}$ to each of the endpoints $(2,-1)$ and $(2,1)$ and the remaining weight $\frac{1}{2}$ to the midpoint $(2,0)$ in the conditional quadratic model. □

The success of this straightforward calculation of the optimum design heavily depends on the block diagonal structure of the information matrix $\mathbf{I}(\delta)$ which vanishes already in slightly more complicated models like the following,

Example 7.2 *One-way layout combined with straight line regression: Common intercept (cf Example 6.5):*

Buonaccorsi (1986) considered the model

$$\widetilde{\mu}(t_1,t_2) = \beta_0 + \beta_1 t_1 + \beta_2 t_2 \ ,$$

$(t_1,t_2) \in \{(t_1,0);\ 0 \le t_1 \le \tau_1\} \cup \{(0,t_2);\ 0 \le t_2 \le \tau_2\}$, where the factors are associated with two kinds of treatments which can be applied alternatively with varying doses t_i and the maximum dose τ_i may be different for the treatments. By means of the Kiefer-Wolfowitz equivalence theorem (Theorem 2.1) it can be checked that the D-optimum design is concentrated on the extreme points $(0,0)$, $(\tau_1,0)$, and $(0,\tau_2)$ and it assigns equal weights $\frac{1}{3}$ to these design points.

As the treatments can be regarded as levels of a qualitative factor the model can be rewritten according to

$$\mu(i,u) = \beta_0 + \beta_i u \ ,$$

$u \in U_i = [0,\tau_i]$, $i = 1,2$, and the interpretation of the parameters remains unchanged. Note that now i is the label of the treatment and u is the dose (cf Example 6.5). □

Motivated by this example models of the form

$$\mu(i,u) = f_0(u)'\beta_0 + f_i(u)'\beta_i \ , \tag{7.3}$$

$u \in U_i$, $i = 1,...,I$, should be the subject of further investigations. There are some tools related to the ideas of Subsections 5.2 and 6.2 which might be helpful for guessing an optimum design. As this is beyond the scope of the present treatise we only add another example for illustrative purposes in which the optimum design unexpectedly shows up to be a product design.

Example 7.3 *One-way layout combined with polynomial regression of differing order: Common intercept:*

As in Example 7.1 we consider a model with a qualitative factor which may be adjusted to two levels and with polynomial regression models associated with these levels which are linear resp. quadratic,

$$\begin{aligned} \mu(1,u) &= \beta_0 + \beta_{11}u \ , \\ \mu(2,u) &= \beta_0 + \beta_{21}u + \beta_{22}u^2 \ , \end{aligned}$$

$u \in U_i$, $i = 1,2$. In contrast to Example 7.1 we have a common constant term β_0 and, hence, the underlying model is of the form (7.3) with $f_0 = \mathbf{1}$, $f_1(u) = u$ and $f_2(u) = (u,u^2)'$. Additionally we assume, again, that the design regions are standardized, $U_1 = U_2 = [-1,1]$.

By means of the KIEFER-WOLFOWITZ equivalence theorem (Theorem 2.1) it can be checked that the D-optimum design assigns equal weights $\frac{1}{4}$ to the extreme points $(1,-1)$, $(1,1)$, $(2,-1)$, and $(2,1)$. This result seems to become more evident if we consider a parametrization

$$\tilde{\mu}(t_1,t_2) = \beta_0 + \beta_1 t_1 + \beta_{2,1}t_2 + \beta_{2,2}t_2^2 \ ,$$

$(t_1,t_2) \in \{(t_1,0);\ -1 \le t_1 \le 1\} \cup \{(0,t_2);\ -1 \le t_2 \le 1\}$ similar to that of Example 7.2 proposed by BUONACCORSI (1986).

Employing FEDOROV's equivalence theorem (Theorem 2.2) we observe that the A-optimum design is also a product design, which assigns weights $\frac{1}{2}(3-\sqrt{6})$ to $(1,-1)$ and $(1,1)$ and weights $\frac{1}{2}(\sqrt{6}-2)$ to $(2,-1)$ and $(2,1)$ respectively. Hence, the marginal design δ_1^* assigns a larger weight $3-\sqrt{6} \approx 0.5505$ to the linear conditional model than to the quadratic conditional model. □

Although the methods presented in this monograph are very powerful tools for the reduction of the dimensionality of the optimization problem involved in the determination of an optimum design they are not omnipotent. As indicated by Example 7.3 there is a number of generally accepted models without a product structure for which different approaches have to be chosen. To give another example we mention briefly a standard model in polynomial regression (cf Examples 4.2, 5.2 and 6.2).

Example 7.4 *Polynomial regression:*
The most common model in multi-dimensional polynomial regression is characterized by a maximum degree M to which the powers of the single factors may add

$$\mu(t) = \sum_{|\alpha| \leq M} t^{\alpha} \beta_{\alpha} ,$$

$t = (t_1, ..., t_K) \in T$, where $\alpha = (\alpha_k)_{k=1,...,K}$ is a multi-index, $\alpha_k \in \{0, 1, 2, ...\}$, $k = 1, ..., K$, $|\alpha| = \sum_{k=1}^{K} \alpha_k$, and $t^{\alpha} = \prod_{k=1}^{K} t_k^{\alpha_k}$. In particular, for $M = 1$ we obtain a linear regression model without interactions $\mu(t_1, ..., t_K) = \beta_0 + \sum_{k=1}^{K} \beta_k t_k$ as treated in Section 5. For $M \geq 2$ the present model differs substantially from those considered so far. For example, in case of two factors ($K = 2$) the polynomial regression up to the second degree ($M = 2$) is given by

$$\mu(t_1, t_2) = \beta_0 + \beta_1 t_1 + \beta_2 t_2 + \beta_{11} t_1^2 + \beta_{22} t_2^2 + \beta_{12} t_1 t_2 .$$

In addition, a variety of design regions T may be regarded as a generalization of the marginal design region of a one-dimensional interval: a standard cube $[-1, 1]^K$, a unit ball $\{t;\ \sum_{k=1}^{K} t_k^2 \leq 1\}$, or a simplex $\{t;\ t_k \geq 0, \sum_{k=1}^{K} t_k = 1\}$ (see FARRELL, KIEFER, and WALBRAN (1967)). Even if we confine to the product-type design region of a standard cube $T = [-1, 1]^K$, generally there will be no product design which is optimum. Also the conditional approach of the present section does not seem to be promising here.

By considerations of invariance with respect to permutations of the factors and by means of the KIEFER-WOLFOWITZ equivalence theorem (Theorem 2.1) D-optimum designs have been obtained by KIEFER (1961) and FARRELL, KIEFER, and WALBRAN (1967) (cf LIM, STUDDEN, and WYNN (1988)). Moreover, LIM and STUDDEN (1988) and DETTE (1990) established that the D-optimum product designs have a high efficiency compared to the actual D-optimum designs. Thus a restriction to product designs is apparently useful also in such situations like the present. □

Further research should include optimum design for unequal variances in multi-factor models (cf WONG (1994)) and the design of large systems (BATES et al. (1995)) including *computer experiments* (cf SACKS et al. (1989) and WELCH et al. (1992)) in which the observations are correlated and the structure of the optimum designs will be completely different (cf MÜLLER-GRONBACH and SCHWABE). (1995)

Concluding remark. In intra-class models optimum design can be generated as a mixture of a family of designs which are conditionally optimum within each class. These ideas have been proposed by KUROTSCHKA (1984) and have been used by BUONACCORSI (1986). Some extensions of this concept are presented.

A Appendix on Partitioned Matrices

In this appendix we mention a few results on partitioned positive-semidefinite symmetric matrices concerning generalized inverses, determinants and identifiability. These results are valuable tools for investigating information matrices.

Let a positive-semidefinite and symmetric matrix $J \in \mathbb{R}^{s \times s}$ be partitioned into submatrices $J_1 \in \mathbb{R}^{s_1 \times s_1}$, $J_2 \in \mathbb{R}^{s_2 \times s_2}$, and $J_{12} \in \mathbb{R}^{s_1 \times s_{21}}$ according to

$$J = \begin{pmatrix} J_1 & J_{12} \\ J_{12}' & J_2 \end{pmatrix}. \tag{A.1}$$

Then J_1 and J_2 are also positive-semidefinite and symmetric.

Lemma A.1

Let J be positive-semidefinite and symmetric. If J is partitioned according to (A.1), then $J_1 J_1^- J_{12} = J_{12}$ for every generalized inverse J_1^- of J_1.

Proof

Because J is positive-semidefinite and symmetrie it can be decomposed according to $J = F'F$ into some $r \times s$ matrix F and its transpose (se e.g. FEDOROV (1972), p 19). Let $F = (F_1|F_2)$ be partitioned into $F_1 \in \mathbb{R}^{s_1 \times r}$ and $F_2 \in \mathbb{R}^{s_2 \times r}$ similarly to J. Then $J_1 = F_1'F_1$, $J_{12} = F_1'F_2$, and the statement follows from the fact that $F_1'F_1(F_1'F_1)^- F_1' = F_1'$ for every generalized inverse of $F_1'F_1$ (see e.g. CHRISTENSEN (1987), p 337). □

Next we turn to the determinant of a partitioned matrix

Lemma A.2

Let J be positive-semidefinite and symmetric. If J is partitioned according to (A.1), then $\det(J) = \det(J_1)\det(J_2 - J_{12}'J_1^- J_{12})$.

Proof

The result follows directly from the decomposition

$$\begin{pmatrix} J_1 & J_{12} \\ J_{12}' & J_2 \end{pmatrix} = \begin{pmatrix} \mathbf{E}_{s_1} & \mathbf{0} \\ J_{12}'J_1^- & \mathbf{E}_{s_2} \end{pmatrix} \begin{pmatrix} J_1 & \mathbf{0} \\ \mathbf{0} & J_2 - J_{12}'J_1^- J_{12} \end{pmatrix} \begin{pmatrix} \mathbf{E}_{s_1} & J_1^- J_{12} \\ \mathbf{0} & \mathbf{E}_{s_2} \end{pmatrix}$$

(see RAGHAVARAO (1971), p 347). □

It is obvious from Lemma A.2 that $\det(J) \le \det(J_1)\det(J_2)$. We finish with a representation of the generalized inverse due to ROHDE (1965).

Lemma A.3

Let J be positive-semidefinite and symmetric. If J is partitioned according to (A.1), then for every generalized inverse J_1^- of J_1

$$J^- = \begin{pmatrix} J_1^- + J_1^- J_{12}(J_2 - J_{12}'J_1^- J_{12})^- J_{12}'J_1^- & -J_1^- J_{12}(J_2 - J_{12}'J_1^- J_{12})^- \\ -(J_2 - J_{12}'J_1^- J_{12})^- J_{12}'J_1^- & (J_2 - J_{12}'J_1^- J_{12})^- \end{pmatrix}$$

is a generalized inverse of J.

Proof

Verification of $JJ^-J = J$ by multiplication. □

References

ATKINSON, A. C. (1982). Developments in the design of experiments. *Int. Statist. Rev.* **50**, 161–177.

ATKINSON, A. C. (1985). An introduction to the optimum design of experiments. In A. C. ATKINSON and S. E. FIENBERG (eds.), *A Celebration of Statistics. The ISI Centenary Volume.* Springer, New York, 465–473.

ATKINSON, A. C. (1988). Recent developments in the methods of optimum and related experimental designs. *Int. Statist. Rev.* **56**, 99–115.

ATKINSON, A. C. (1995). The usefulness of optimum experimental designs (with discussion). *J. R. Statist. Soc., Ser. B* **57** (to appear).

ATKINSON, A. C. and A. N. DONEV (1992). *Optimum Experimental Designs.* Clarendon Press, Oxford.

ATWOOD, C. L. (1969). Optimal and efficient designs of experiments. *Ann. Math. Statist.* **40**, 1570–1602.

ATWOOD, C. L. (1973). Sequences converging to *D*-optimal designs of experiments. *Ann. Statist.* **1**, 342–352.

BAKSALARY, J. K. (1984). A study of the equivalence between a Gauss-Markoff model and its augmentation by nuisance parameters. *Math. Operationsforsch. Statist., ser. statist.* **15**, 3–35.

BANDEMER, H. et al. (1977). *Theorie und Anwendung der optimalen Versuchsplanung I. Handbuch zur Theorie.* Akademie-Verlag, Berlin.

BANDEMER, H. and A. BELLMANN (1994). *Statistische Versuchsplanung. 4th ed.* Teubner, Stuttgart.

BANDEMER, H. and W. NÄTHER (1980). *Theorie und Anwendung der optimalen Versuchsplanung II. Handbuch zur Anwendung.* Akademie-Verlag, Berlin.

BASSO, L., A. WINTERBOTTOM and H. P. WYNN (1986). A review of the "Taguchi methods" for off-line quality control. *Quality Reliability Eng. Int.* **2**, 71–79.

BATES, R. A., R. J. BUCK, E. RICCOMAGNO and H. P. WYNN (1995). Experimental design and observations for large systems (with discussion). *J. R. Statist. Soc., Ser. B* **57** (to appear).

BILLINGSLEY, P. (1968). *Convergence of Probability Measures.* Wiley, New York.

BUONACCORSI J. P. (1986). Designs for slope ratio assays. *Biometrics* **42**, 875–882.

BUONACCORSI J. P. and H. K. IYER (1986). Optimal designs for ratios of linear combinations in the general linear model. *J. Statist. Plann. Inference* **13**, 345–356.

CHENG, C. S. (1988). Biased weighing designs. Universität Augsburg, "Anwendungsbezogene Optimierung und Steuerung", *Report No 59.*

CHRISTENSEN, R. (1987). *Plane Answers to Complex Questions: The Theory of Linear Models* Springer, New York.

COOK, D. R. and L. A. THIBODEAU (1980). Marginally restricted D-optimal designs. *J. Amer. Statist. Assoc.* **75**, 366–371.

COX, D. R. (1984). Present position and potential developments: Some personal views - Design of experiments and regression. *J. R. Statist. Soc., Ser. A* **147**, 306–315.

DAVID, O. (1994). Optimal designs for comparing several treatments with a control in the presence of explanatory variables. *Statistics* **25**, 325–331.

DETTE, H. (1990). A generalization of D- and D_1-optimal designs in polynomial regression. *Ann. Statist.* **18**, 1784–1804.

DONEV, A. N. (1989). Design of experiments with both mixture and qualitative factors. *J. R. Statist. Soc., Ser. B* **50**, 297–302.

DRAPER, N. R. and J. A. JOHN (1988). Response surface designs for quantitative and qualitative variables. *Technometrics* **30**, 423–428.

EATON, M. L. (1989). *Group Invariance Applications in Statistics.* em Regional Conference Series in Probability and Statistics **1**. Institute of Mathematical Statistics, Hayward.

EHRENFELD, S. (1956). Complete class theorems in experimental design. *Proc. Third Berkeley Sympos. Math. Statist. Probab.*, **1**, 57–67.

ELFVING, G. (1952). Optimum allocation in linear regression theory. *Ann. Math. Statist.* **23**, 255–262.

ERMAKOV, S. M. et al. (1983). *Mathematical Theory of Experimental Design.* Nauka, Moscow (in Russian).

ERMAKOV, S. M. and V. B. MELAS (1995). *Design and Analysis of Simulation Experiments.* Kluwer, Dordrecht.

ERMAKOV, S. M. and A. A. ZHIGLYAVSKY (1987). *Mathematical Theory of Optimal Experiments.* Nauka, Moscow (in Russian).

FARRELL, R. H., J. KIEFER and A. WALBRAN (1967). Optimum multivariate designs. *Proc. Fifth Berkeley Sympos. Math. Statist. Probab.*, **1**, 113–138.

FEDOROV, V. V. (1971). Design of experiments for linear optimality criteria. *Theory Probab. Appl.* **16**, 189–195.

FEDOROV, V. V. (1972). *Theory of Optimal Experiments.* Academic Press, New York.

FISHER, R. (1935). *The Design of Experiments.* Oliver Boyd, London.

GAFFKE, N. (1981). Some classes of optimality criteria and optimal designs for complete two-way layouts. *Ann. Statist.* **9**, 893–898.

GAFFKE, N. (1987). Further characterizations of design optimality and admissibility for partial parameter estimation in linear regression. *Ann. Statist.* **15**, 942–957.

GAFFKE, N. and B. HEILIGERS (1995). Approximate designs for linear regression: Invariance, admissibility, and optimality. *Otto-von-Guericke-Universität Magdeburg, Fakultät für Mathematik, Preprint Nr. 12, 1995.*

GAFFKE, N. and O. KRAFFT (1979). Optimum designs in complete two-way layouts. *J. Statist. Plann. Inference* **3**, 119–126.

GIOVAGNOLI, A., F. PUKELSHEIM and H. P. WYNN (1987). Group invariant orderings and experimental designs. *J. Statist. Plann. Inference* **17**, 111–135.

GRAHAM, A. (1981). *Kronecker Products and Matrix Calculus: with Applications.* Ellis Horwood, Chichester.

GUEST, P. G. (1958). The spacing of observations in polynomial regression. *Ann. Math. Statist.* **29**, 294–299.

HALMOS, P. R. (1974). *Measure Theory.* Springer, New York.

HARVILLE, D. A. (1975). Computing optimum designs for covariance models. In J. N. SRIVASTAVA(ed.), *A Survey of Statistical Design and Linear Models.* North-Holland, Amsterdam, 209–228.

HOEL, P. G. (1958). Efficiency problems in polynomial estimation. *Ann. Math. Statist.* **29**, 1134–1145.

HOEL, P. G. (1965). Minimax designs in two dimensional regression. *Ann. Math. Statist.* **36**, 1097–1106.

HUANG, M.-N. L. and M.-C. HSU (1993). Marginally restricted linear-optimal designs. *J. Statist. Plann. Inference* **35**, 251–266.

HUDA, S. and R. MUKERJEE (1988). Optimal weighing designs: Approximate theory. *statistics* **19**, 513–517.

KARLIN, S. and W. J. STUDDEN (1966). Optimal experimental designs. *Ann. Math. Statist.* **37**, 783–815.

KIEFER, J. (1959). Optimum experimental designs. *J. R. Statist. Soc., Ser. B* **21**, 272–304.

KIEFER, J. (1961). Optimum designs in regression problems, II. *Ann. Math. Statist.* **32**, 298–325.

KIEFER, J. (1974). General equivalence theory for optimum designs (approximate theory). *Ann. Statist.* **2**, 849–879.

KIEFER, J. (1985). *Collected Papers III. Design of Experiments.* Springer, New York.

KIEFER, J. and J. WOLFOWITZ (1959). Optimum designs in regression problems. *Ann. Math. Statist.* **30**, 271–294.

KIEFER, J. and J. WOLFOWITZ (1960). The equivalence of two extremum problems. *Can. J. Math.* **12**, 363–366.

KRAFFT, O. (1978). *Lineare statistische Modelle und optimale Versuchspläne.* Vandenhoeck & Ruprecht, Göttingen.

KUNERT, J. (1983). Optimal design and refinement of the linear model with applications to repeated measurement designs. *Ann. Statist.* **11**, 247–257.

KUROTSCHKA, V. (1971). Optimale Versuchspläne bei zweifach klassifizierten Beobachtungsmodellen. *Metrika* **17**, 215–232.

KUROTSCHKA, V. (1972). *Optimale Versuchsplanung bei Modellen der Varianzanalyse.* Habilitationsschrift. Universität Göttingen.

KUROTSCHKA, V. (1978). Optimal design of complex experiments with qualitative factors of influence. *Commun. Statist., Theory Methods* **A7**, 1363–1378.

KUROTSCHKA, V. G. (1984). A general approach to optimum design of experiments with qualitative and quantitative factors. In J. K. GHOSH and J. ROY (eds.), *Statistics: Applications and New Directions: Proceedings of the Indian Statistical Institute Golden Jubilee International Conference Calcutta 1981*, 353–368.

KUROTSCHKA, V. G. (1985). llustrazione del disegno ottimale degli esperimenti generali di locazione. In *Rassegna di Metodi Statistici ed Applicazioni 5, Cagliari 1985.* Pitagora, Bologna, 33–70.

KUROTSCHKA, V. G. (1987). Optimum design of general intraclass regression experiments and general analysis of covariance experiments. In *Proc. Fourth Vilnius Conf. 1985, Vol. 2.* VNU Science Press, Utrecht, 167–171.

KUROTSCHKA, V. G. (1988). Characterizations and examples of optimal experiments with qualitative and quantitative factors. In V. FEDOROV and H. LÄUTER (eds.), *Model-Oriented Data Analysis, Proceedings Eisenach 1987.* Springer, Berlin, 53–71.

KUROTSCHKA, V. G. and R. SCHWABE (1995). The reduction of design problems for multivariate experiments to univariate possibilities and their limitations. *Freie Universität Berlin, Fachbereich Mathematik und Informatik, Preprint No. A/33/95.*

KUROTSCHKA, V., R. SCHWABE and W. WIERICH (1992). Optimum designs for partly interacting qualitative and quantitative factors. In A. PÁZMAN and J. VOLAUFOVÁ (eds.), *Probastat '91, Proc. Int. Conf. Probab. Math. Statist., Bratislava 1991.* Slovak Academy of Sciences, Bratislava, 102–108.

KUROTSCHKA, V. and W. WIERICH (1984a). Optimale Planung eines Kovarianzanalyse- und eines Intraclass Regressions-Experiments. *Metrika* **31**, 361–378.

KUROTSCHKA, V. and W. WIERICH (1984b). Ansätze und Ergebnisse der optimalen Planung von Experimenten mit sowohl qualitativen als auch quantitativen Einflussfaktoren. In M. IOSIFESCU (ed.), *Proc. Seventh Conf. Probab. Theory, Braşov 1982.* Editura, Bucharest, 237–248.

LÄUTER, E. (1974). Experimental design in a class of models. *Math. Operationsforsch. Statist.* **5**, 379–398.

LEHMANN, E. L. (1959). *Testing Statistical Hypothesis.* Wiley, New York.

LIM, Y. B. and W. J. STUDDEN (1988). Efficient D_s-optimal designs for multivariate polynomial regression on the q-cube. *Ann. Statist.* **16**, 1225–1240.

LIM, Y. B., W. J. STUDDEN, and H. P. WYNN (1988). A note on approximate D-optimal designs for $G \times 2^M$. In S. S. GUPTA and J. O. BERGER (eds.), *Statistical Decision Theory and Related Topics IV, Vol. 2.* Springer, New York, 351–361.

LOGOTHETIS, N. and H. P. WYNN (1989). *Quality Through Design. Experimental Design, Off-line Quality Control and Taguchi's Contributions.* Clarendon Press, Oxford.

LOPES TROYA, J. (1982a). Optimal designs for covariates' models. *J. Statist. Plann. Inference* **6**, 373–419.

LOPES TROYA, J. (1982b). Cyclic designs for a covariate model. *J. Statist. Plann. Inference* **7**, 49–75.

MAGDA, C. G. (1980). Circular balanced repeated measurement designs. *Commun. Statist., Theory Methods* **A9**, 1901–1918.

MAGNUS, J. R. and H. NEUDECKER (1988). *Matrix Differential Calculus with Applications in Statistics and Econometrics.* Wiley, Chichester.

MELAS, V. B. (1982). A duality theorem for E-optimality. *Zavodskaya Laboratoria* **48**, 48–50 (in Russian).

MÜLLER-GRONBACH, T. and R. SCHWABE (1995). On optimal allocations for estimating the surface of a random field. *Metrika* (to appear).

NÄTHER, W. and V. REINSCH (1981). D_s-optimality and WHITTLE's equivalence theorem. *Math. Operationsforsch. Statist., ser. statist.* **12**, 307–316.

PÁZMAN, A. (1980). Singular experimental designs (standard and HILBERT-space approaches). *Math. Operationsforsch. Statist., ser. statist.* **11**, 137–149.

PÁZMAN, A. (1986). *Foundations of Optimum Experimental Design.* Reidel, Dordrecht.

PETERSEN, I. F. and Y. P. KUKS (1971). Direct product methods for the construction of plans of regressive experiments. *Industrial Laboratory* **37**, 74–77.

PUKELSHEIM, F. (1980). On linear regression designs which maximize information. *J. Statist. Plann. Inference* **4**, 339–364.

PUKELSHEIM, F. (1981). On *c*-optimal design measures. *Math. Operationsforsch. Statist., ser. statist.* **12**, 13–20.

PUKELSHEIM, F. (1983). On optimality properties of simple block designs in the approximate design theory. *J. Statist. Plann. Inference* **8**, 193–208.

PUKELSHEIM, F. (1987). Information increasing orderings in experimental design theory. *Int. Statist. Rev.* **55**, 203–219.

PUKELSHEIM, F. (1993). *Optimal Design of Experiments.* Wiley, New York.

PUKELSHEIM, F. and D. M. TITTERINGTON (1983). General differential and Lagrangian theory for optimal experimental design. *Ann. Statist.* **11**, 1060–1068.

RAFAJŁOWICZ, E. and W. MYSZKA (1988). Optimum experimental design for a regression on a hypercube - generalization of Hoel's result. *Ann. Inst. Statist. Math.* **40**, 821–827.

RAFAJŁOWICZ, E. and W. MYSZKA (1992). When product type experimental design is optimal? Brief survey and new results. *Metrika* **39**, 321–333.

RAGHAVARAO, D. (1971). *Constructions and Combinatorial Problems in Design of Experiments.* Wiley, New York.

RAKTOE, B. L., A. HEDAYAT and W. T. FEDERER (1981). *Factorial Designs.* Wiley, New York.

RAO, C. R. (1973). *Linear Statistical Inference and Its Applications, 2nd ed.* Wiley, New York.

RAO, C. R. and S. K. MITRA (1971). *Generalized Inverse of Matrices and Its Applications.* Wiley, New York.

RICCOMAGNO, E., R. SCHWABE and H. P. WYNN (1995). Lattice-based optimum design for Fourier regression. *Freie Universität Berlin, Fachbereich Mathematik und Informatik, Preprint No. A/15/95.*

ROHDE, C. A. (1965). Generalized inverses of partitioned matrices. *J. Soc. Indust. Appl. Math.* **13**, 1033–1035.

SACKS, J., W. J. WELCH, T. J. MITCHELL and H. P. WYNN (1989). Design and analysis of computer experiments. *Statist. Science* **4**, 409–435.

SCHWABE, R. (1995a). Experimental design for linear models with higher order interaction terms. In V. MAMMITZSCH and H. SCHNEEWEISS (eds.), *Symposia Gaussiana. Proceedings of the 2nd Gauss Symposium, München 1993. Conference B: Statistical Sciences* DeGruyter, Berlin, 281–288.

SCHWABE, R. (1995b). Optimal designs for additive linear models. *Statistics* (to appear).

SCHWABE, R. (1995c). Model robust experimental design in the presence of interactions: The orthogonal case. In A. PÁZMAN (ed.), *PROBASTAT'94. Proceedings of the International Conference on Probability and Mathematical Statistics, Smolenice 1994. Tatra Mountains Math. Publ.* **7** (to appear).

SCHWABE, R. (1995d). Optimal designs for complete interaction structures. *Freie Universität Berlin, Fachbereich Mathematik und Informatik, Preprint No. A/25/95.*

SCHWABE, R. and W. WIERICH (1995). *D*-optimal designs of experiments with non-interacting factors. *J. Statist. Plann. Inference* **44**, 371–384.

SCHWABE, R. and W. K. WONG (1995). Some relationships between efficiencies and marginal efficiencies of product designs. *Freie Universität Berlin, Fachbereich Mathematik und Informatik, Preprint No. A/18/95.*

SEARLE, S. R. (1971). *Linear Models.* Wiley, New York.

SHAH, K. R. and B. K. SINHA (1989). *Theory of Optimal Designs. Lecture Notes in Statistics* **54**. Springer, New York.

SILVEY, S. D. (1978). Optimal design measures with singular information matrices. *Biometrika* **65**, 553–559.

SILVEY, S. D. (1980). *Optimal Design.* Chapman and Hall, London.

SMITH, K. (1918). On the standard deviation of adjusted and interpolated values of an observed *polynomial function* and its constants and the guidance towards a proper choice of the distribution of observations. *Biometrika* **12**, 1–85.

STUDDEN, W. J. (1980). D_s-optimal designs for polynomial regression using continued fractions. *Ann. Statist.* **8**, 1132–1141.

TAGUCHI, G. (1987). *System of Experimental Design. Vols. 1 and 2.* UNIPUB/Kraus International, White Plains.

VUCHKOV, I. N., C. A. YONTCHEV, and D. L. DAMGALIEV (1983). Continuous D-optimal designs for experiments with mixture and process variables. *Math. Operationsforsch. Statist., ser. statist.* **14**, 33–51.

WALD, A. (1943). On efficient design of statistical investigations. *Ann. Math. Statist.* **14**, 134–140.

WELCH, W. J., R. J. BUCK, J. SACKS, H. P. WYNN, T. J. MITCHELL, and M. D. MORRIS (1992). Screening, predicting, and computer experiments. *Technometrics* **34**, 15–25.

WHITTLE, P. (1973). Some general points in the theory of optimal experimental design. *J. R. Statist. Soc., Ser. B* **35**, 123–130.

WIERICH, W. (1984). Konkrete optimale Versuchspläne für ein lineares Modell mit einem qualitativen und zwei quantitativen Einflußfaktoren. *Metrika* **31**, 285–301.

WIERICH, W. (1985). Optimum designs under experimental constraints for a covariate model and an intra-class regression model. *J. Statist. Plann. Inference* **12**, 27–40.

WIERICH, W. (1986a). Optimality of k-proportional designs for simultaneous inference. *statistics* **17**, 179–187.

WIERICH, W. (1986b). On optimal designs and complete class theorems for experiments with continuous and discrete factors of influence. *J. Statist. Plann. Inference* **15**, 19–27.

WIERICH, W. (1986c). *The D- and A-optimality of product design measures for linear models with discrete and continuous factors of influence.* Habilitationsschrift. Freie Universität Berlin, Fachbereich Mathematik.

WIERICH, W. (1988a). A-optimal design measures for one-way layouts with additive regression. *J. Statist. Plann. Inference* **18**, 57–68.

WIERICH, W. (1988b). On A-optimality and non-optimality of the best proportional design measures in ANOVA-settings. *statistics* **19**, 503–512.

WIERICH, W. (1989a). An application of the Kiefer-Wolfowitz equivalence theorem to ANOVA models with additive regression. *J. Statist. Plann. Inference* **21**, 277–283.

WIERICH, W. (1989b). On the relative efficiency of the best second order proportional ANOVA-designs. *Comput. Statist. Data Anal.* **8**, 119–134.

WIJSMAN, R. A. (1990). *Invariant Measures on Groups and Their Use in Statistics. IMS Lecture Notes - Monograph Series* **14**. Institute of Mathematical Statistics, Hayward.

WONG, W. K. (1994). G-optimal designs for multi-factor experiments with heteroscedastic errors. *J. Statist. Plann. Inference* **40**, 127–133.

WYNN, H. P. (1984). Jack Kiefer's work on experimental design. *Ann. Statist.* **12**, 716–723.

List of Symbols

a	regression function
$a^{(k)}$	regression function in the kth marginal model
β	vector uf unknown parameters
$\beta^{(k)}$	vector uf unknown parameters in the kth marginal model
$\mathbf{C}(\delta)$	$= \mathbf{I}(\delta)^{-1}$ covariance matrix of the design $\delta \in \Delta(\beta)$
$\mathbf{C}_k(\delta_k)$	$= \mathbf{I}_k(\delta_k)^{-1}$ covariance matrix of the design $\delta_k \in \Delta_k(\beta^{(k)})$ in the kth marginal model
$\mathbf{C}_\psi(\delta)$	$= L_\psi \mathbf{I}(\delta)^- L'_\psi$ covariance matrix of the design $\delta \in \Delta(\psi)$ for ψ
$\mathbf{C}_{k,\psi_k}(\delta_k)$	$= L_{k,\psi_k} \mathbf{I}_k(\delta_k)^- L'_{k,\psi_k}$ covariance matrix of the design $\delta_k \in \Delta_k(\psi_k)$ for ψ_k in the kth marginal model
δ	design
δ_k	design in the kth marginal model
$\overline{\delta}$	invariant design
$\overline{\delta}_k$	invariant design in the kth marginal model
Δ	set of designs on T
Δ_k	set of designs on T_k in the kth marginal model
$\Delta(\psi)$	set of designs for which ψ is identifiable
$\Delta_k(\psi_k)$	set of designs for which ψ_k is identifiable in the kth marginal model
$\Delta(\beta)$	set of designs for which β is identifiable
$\Delta_k(\beta^{(k)})$	set of designs for which $\beta^{(k)}$ is identifiable in the kth marginal model
$\det(\cdot)$	determinant of a matrix
$\text{diag}(\cdot)$	diagonal or block diagonal matrix
$\mathbf{e}_{m,n}$	$m \times 1$ vector, nth entry equal to one, all other entries equal to zero ($\mathbf{e}_{m,n} = \mathbf{0}$ if $n \notin \{1, ..., m\}$)
$\mathbf{E}_n$	$n \times n$ identity matrix
ϵ_t	one-point design (DIRAC-measure) at t
g	transformations on T
g_k	transformations on T_k in the kth marginal model
G	group of transformations on T
G_k	group of transformations on T_k in the kth marginal model
$\mathbf{I}(\delta)$	information matrix of the design δ

$\mathbf{I}_k(\delta_k)$	information matrix of the design δ_k in the kth marginal model
K	number of factors of influence
L_ψ	selection matrix corresponding to ψ
L_{k,ψ_k}	selection matrix corresponding to ψ_k in the kth marginal model
$\lambda_n(\cdot)$	eigenvalue
$\lambda_{\max}(\cdot)$	maximal eigenvalue
$\lambda_{\min}(\cdot)$	minimal eigenvalue
μ	response function
μ_k	response function in the kth marginal model
∇_φ	derivative of φ
∇_{φ_k}	derivative of φ_k in the kth marginal model
p	dimension of β
p_k	dimension of $\beta^{(k)}$ in the kth marginal model
Φ	criterion function on δ
$\Phi^{(k)}$	criterion function on δ_k in the kth marginal model
ϕ	criterion function on $\mathbf{I}(\delta)$
φ	criterion function on $\mathbf{C}_\psi(\delta)$
φ_k	criterion function on $\mathbf{C}_{k,\psi_k}(\delta_k)$ in the kth marginal model
ψ	linear aspect
ψ_k	linear aspect in the kth marginal model
t	design point
t_k	design point in the kth marginal model
T	design region
T_k	design region in the kth marginal model
$\mathrm{tr}(\cdot)$	trace of a matrix
$\mathbf{0}$	vector or matrix of zeroes
$\mathbf{1}$	constant function, equal to one
$\mathbf{1}_n$	$n \times 1$ vector with all entries equal to one
$\mathbf{1}_m^n$	$m \times n$ matrix with all entries equal to one
$\cdot^-$	generalized inverse of a matrix: $MM^-M = M$
$\cdot \leq \cdot$	semi-ordering of positive-semidefinite matrices: $M_1 \leq M_2$ if $M_2 - M_1$ is positive-semidefinite
$\cdot \otimes \cdot$	– KRONECKER-product of matrices (resp. vectors): $M_1 \otimes M_2$ – (measure theoretical) product of designs: $\delta_1 \otimes \delta_2$

Index

Lecture Notes in Statistics

For information about Volumes 1 to 29 please contact Springer-Verlag

Vol 30: J. Grandell, Stochastic Models of Air Pollutant Concentration. v, 110 pages, 1985.

Vol. 31: J. Pfanzagl, Asymptotic Expansions for General Statistical Models. vii, 505 pages, 1985.

Vol. 32: R. Gilchrist, B. Francis and J. Whittaker (Editors), Generalized Linear Models. Proceedings, 1985. vi, 178 pages, 1985.

Vol. 33: M. Csorgo, S. Csorgo, L. Horvath, An Asymptotic Theory for Empirical Reliability and Concentration Processes. v, 171 pages, 1986.

Vol. 34: D.E. Critchlow, Metric Methods for Analyzing Partially Ranked Data. x, 216 pages, 1985.

Vol. 35: T. Calinski and W. Klonecki (Editors), Linear Statistical Inference. Proceedings, 1984. vi, 318 pages, 1985.

Vol. 36: B. Matern, Spatial Variation. Second Edition. 151 pages, 1986.

Vol. 37: R. Dykstra, T. Robertson and F.T. Wright (Editors), Advances in Order Restricted Statistical Inference. Proceedings, 1985. viii, 295 pages, 1986.

Vol. 38: R.W. Pearson and R.F. Boruch (Editors), Survey Research Designs: Towards a Better Understanding of Their Costs and Benefits. v, 129 pages, 1986.

Vol. 39: J.D. Malley, Optimal Unbiased Estimation of Variance Components. ix, 146 pages, 1986.

Vol. 40: H.R. Lerche, Boundary Crossing of Brownian Motion. v, 142 pages, 1986.

Vol. 41: F. Baccelli, P. Bremaud, Palm Probabilities and Stationary Queues. vii, 106 pages, 1987.

Vol. 42: S. Kullback, J.C. Keegel, J.H. Kullback, Topics in Statistical Information Theory. ix, 158 pages, 1987.

Vol. 43: B.C. Arnold, Majorization and the Lorenz Order: A Brief Introduction. vi, 122 pages, 1987.

Vol. 44: D.L. McLeish, Christopher G. Small, The Theory and Applications of Statistical Inference Functions. vi, 124 pages, 1987.

Vol. 45: J.K. Ghosh (Editor), Statistical Information and Likelihood. 384 pages, 1988.

Vol. 46: H.-G. Muller, Nonparametric Regression Analysis of Longitudinal Data. vi, 199 pages, 1988.

Vol. 47: A.J. Getson, F.C. Hsuan, {2}-Inverses and Their Statistical Application. viii, 110 pages, 1988.

Vol. 48: G.L. Bretthorst, Bayesian Spectrum Analysis and Parameter Estimation. xii, 209 pages, 1988.

Vol. 49: S.L. Lauritzen, Extremal Families and Systems of Sufficient Statistics. xv, 268 pages, 1988.

Vol. 50: O.E. Barndorff-Nielsen, Parametric Statistical Models and Likelihood. vii, 276 pages, 1988.

Vol. 51: J. Husler, R.-D. Reiss (Editors). Extreme Value Theory, Proceedings, 1987. x, 279 pages, 1989.

Vol. 52: P.K. Goel, T. Ramalingam, The Matching Methodology: Some Statistical Properties. viii, 152 pages, 1989.

Vol. 53: B.C. Arnold, N. Balakrishnan, Relations, Bounds and Approximations for Order Statistics. ix, 173 pages, 1989.

Vol. 54: K.R. Shah, B.K. Sinha, Theory of Optimal Designs. viii, 171 pages, 1989.

Vol. 55: L. McDonald, B. Manly, J. Lockwood, J. Logan (Editors), Estimation and Analysis of Insect Populations. Proceedings, 1988. xiv, 492 pages, 1989.

Vol. 56: J.K. Lindsey, The Analysis of Categorical Data Using GLIM. v, 168 pages, 1989.

Vol. 57: A. Decarli, B.J. Francis, R. Gilchrist, G.U.H. Seeber (Editors), Statistical Modelling. Proceedings, 1989. ix, 343 pages, 1989.

Vol. 58: O.E. Barndorff-Nielsen, P. Blaesild, P.S. Eriksen, Decomposition and Invariance of Measures, and Statistical Transformation Models. v, 147 pages, 1989.

Vol. 59: S. Gupta, R. Mukerjee, A Calculus for Factorial Arrangements. vi, 126 pages, 1989.

Vol. 60: L. Gyorfi, W. Hardle, P. Sarda, Ph. Vieu, Nonparametric Curve Estimation from Time Series. viii, 153 pages, 1989.

Vol. 61: J. Breckling, The Analysis of Directional Time Series: Applications to Wind Speed and Direction. viii, 238 pages, 1989.

Vol. 62: J.C. Akkerboom, Testing Problems with Linear or Angular Inequality Constraints. xii, 291 pages, 1990.

Vol. 63: J. Pfanzagl, Estimation in Semiparametric Models: Some Recent Developments. iii, 112 pages, 1990.

Vol. 64: S. Gabler, Minimax Solutions in Sampling from Finite Populations. v, 132 pages, 1990.

Vol. 65: A. Janssen, D.M. Mason, Non-Standard Rank Tests. vi, 252 pages, 1990.

Vol 66: T. Wright, Exact Confidence Bounds when Sampling from Small Finite Universes. xvi, 431 pages, 1991.

Vol. 67: M.A. Tanner, Tools for Statistical Inference: Observed Data and Data Augmentation Methods. vi, 110 pages, 1991.

Vol. 68: M. Taniguchi, Higher Order Asymptotic Theory for Time Series Analysis. viii, 160 pages, 1991.

Vol. 69: N.J.D. Nagelkerke, Maximum Likelihood Estimation of Functional Relationships. V, 110 pages, 1992.

Vol. 70: K. Iida, Studies on the Optimal Search Plan. viii, 130 pages, 1992.

Vol. 71: E.M.R.A. Engel, A Road to Randomness in Physical Systems. ix, 155 pages, 1992.

Vol. 72: J.K. Lindsey, The Analysis of Stochastic Processes using GLIM. vi, 294 pages, 1992.

Vol. 73: B.C. Arnold, E. Castillo, J.-M. Sarabia, Conditionally Specified Distributions. xiii, 151 pages, 1992.

Vol. 74: P. Barone, A. Frigessi, M. Piccioni, Stochastic Models, Statistical Methods, and Algorithms in Image Analysis. vi, 258 pages, 1992.

Vol. 75: P.K. Goel, N.S. Iyengar (Eds.), Bayesian Analysis in Statistics and Econometrics. xi, 410 pages, 1992.

Vol. 76: L. Bondesson, Generalized Gamma Convolutions and Related Classes of Distributions and Densities. viii, 173 pages, 1992.

Vol. 77: E. Mammen, When Does Bootstrap Work? Asymptotic Results and Simulations. vi, 196 pages, 1992.

Vol. 78: L. Fahrmeir, B. Francis, R. Gilchrist, G. Tutz (Eds.), Advances in GLIM and Statistical Modelling: Proceedings of the GLIM92 Conference and the 7th International Workshop on Statistical Modelling, Munich, 13-17 July 1992. ix, 225 pages, 1992.

Vol. 79: N. Schmitz, Optimal Sequentially Planned Decision Procedures. xii, 209 pages, 1992.

Vol. 80: M. Fligner, J. Verducci (Eds.), Probability Models and Statistical Analyses for Ranking Data. xxii, 306 pages, 1992.

Vol. 81: P. Spirtes, C. Glymour, R. Scheines, Causation, Prediction, and Search. xxiii, 526 pages, 1993.

Vol. 82: A. Korostelev and A. Tsybakov, Minimax Theory of Image Reconstruction. xii, 268 pages, 1993.

Vol. 83: C. Gatsonis, J. Hodges, R. Kass, N. Singpurwalla (Editors), Case Studies in Bayesian Statistics. xii, 437 pages, 1993.

Vol. 84: S. Yamada, Pivotal Measures in Statistical Experiments and Sufficiency. vii, 129 pages, 1994.

Vol. 85: P. Doukhan, Mixing: Properties and Examples. xi, 142 pages, 1994.

Vol. 86: W. Vach, Logistic Regression with Missing Values in the Covariates. xi, 139 pages, 1994.

Vol. 87: J. Moller, Lectures on Random Voronoi Tessellations.vii, 134 pages, 1994.

Vol. 88: J. E. Kolassa, Series Approximation Methods in Statistics.viii, 150 pages, 1994.

Vol. 89: P. Cheeseman, R.W. Oldford (Editors), Selecting Models From Data: AI and Statistics IV. xii, 487 pages, 1994.

Vol. 90: A. Csenki, Dependability for Systems with a Partitioned State Space: Markov and Semi-Markov Theory and Computational Implementation. x, 241 pages, 1994.

Vol. 91: J.D. Malley, Statistical Applications of Jordan Algebras. viii, 101 pages, 1994.

Vol. 92: M. Eerola, Probabilistic Causality in Longitudinal Studies. vii, 133 pages, 1994.

Vol. 93: Bernard Van Cutsem (Editor), Classification and Dissimilarity Analysis. xiv, 238 pages, 1994.

Vol. 94: Jane F. Gentleman and G.A. Whitmore (Editors), Case Studies in Data Analysis. viii, 262 pages, 1994.

Vol. 95: Shelemyahu Zacks, Stochastic Visibility in Random Fields. x, 175 pages, 1994.

Vol. 96: Ibrahim Rahimov, Random Sums and Branching Stochastic Processes. viii, 195 pages, 1995.

Vol. 97: R. Szekli, Stochastic Ordering and Dependence in Applied Probability. viii, 194 pages, 1995.

Vol. 98: Philippe Barbe and Patrice Bertail, The Weighted Bootstrap. viii, 230 pages, 1995.

Vol. 99: C.C. Heyde (Editor), Branching Processes: Proceedings of the First World Congress. viii, 185 pages, 1995.

Vol. 100: Wlodzimierz Bryc, The Normal Distribution: Characterizations with Applications. viii, 139 pages, 1995.

Vol. 101: H.H. Andersen, M.Hojbjerre, D. Sorensen, P.S.Eriksen, Linear and Graphical Models: for the Multivariate Complex Normal Distribution. x, 184 pages, 1995.

Vol. 102: A.M. Mathai, Serge B. Provost, Takesi Hayakawa, Bilinear Forms and Zonal Polynomials. x, 378 pages, 1995.

Vol. 103: Anestis Antoniadis and Georges Oppenheim (Editors), Wavelets and Statistics. vi, 411 pages, 1995.

Vol. 104: Gilg U.H. Seeber, Brian J. Francis, Reinhold Hatzinger, Gabriele Steckel-Berger (Editors), Statistical Modelling: 10th International Workshop, Innsbruck, July 10-14th 1995. x, 327 pages, 1995.

Vol. 105: Constantine Gatsonis, James S. Hodges, Robert E. Kass, Nozer D. Singpurwalla(Editors), Case Studies in Bayesian Statistics, Volume II. x, 354 pages, 1995.

Vol. 106: Harald Niederreiter, Peter Jau-Shyong Shiue (Editors), Monte Carlo and Quasi-Monte Carlo Methods in Scientific Computing. xiv, 372 pages, 1995.

Vol. 107: Masafumi Akahira, Kei Takeuchi, Non-Regular Statistical Estimation. vii, 183 pages, 1995.

Vol. 108: Wesley L. Schaible (Editor), Indirect Estimators in U.S. Federal Programs. iix, 195 pages, 1995.

Vol. 109: Helmut Rieder (Editor), Robust Statistics, Data Analysis, and Computer Intensive Methods. xiv, 427 pages, 1996.

Vol. 110: D. Bosq, Nonparametric Statistics for Stochastic Processes. xii, 169 pages, 1996.

Vol. 111: Leon Willenborg, Ton de Waal, Statistical Disclosure Control in Practice. xiv, 152 pages, 1996.

Vol. 112: Doug Fischer, Hans-J. Lenz (Editors), Learning from Data. xii, 450 pages, 1996.

Vol. 113: Rainer Schwabe, Optimum Designs for Multi-Factor Models. viii, 124 pages, 1996.